Graph Theory in Operations Research

Macmillan Computer Science Series

Consulting Editor
Professor F. H. Sumner, University of Manchester

S. T. Allworth, *Introduction to Real-time Software Design*

Ian O. Angell, *A Practical Introduction to Computer Graphics*

G. M. Birtwistle, *Discrete Event Modelling on Simula*

T. B. Boffey, *Graph Theory in Operations Research*

Richard Bornat, *Understanding and Writing Compilers*

J. K. Buckle, *The ICL 2900 Series*

Derek Coleman, *A Structured Programming Approach to Data**

Andrew J. T. Colin, *Fundamentals of Computer Science*

Andrew J. T. Colin, *Programming and Problem-solving in Algol 68**

S. M. Deen, *Fundamentals of Data Base Systems**

J. B. Gosling, *Design of Arithmetic Units for Digital Computers*

David Hopkin and Barbara Moss, *Automata**

Roger Hutty, *Fortran for Students*

H. Kopetz, *Software Reliability*

A. M. Lister, *Fundamentals of Operating Systems, second edition**

G. P. McKeown and V. J. Rayward-Smith, *Mathematics for Computing*

Brian Meek, *Fortran, PL/I and the Algols*

Derrick Morris and Roland N. Ibbett, *The MU5 Computer System*

John Race, *Case Studies in Systems Analysis*

B. S. Walker, *Understanding Microprocessors*

I. R. Wilson and A. M. Addyman, *A Practical Introduction to Pascal*

* The titles marked with an asterisk were prepared during the Consulting Editorship of Professor J. S. Rohl, University of Western Australia.

Graph Theory in Operations Research

T. B. Boffey
University of Liverpool

© T. B. Boffey 1982

All rights reserved. No part of this publication may be reproduced or transmitted, in any form or by any means, without permission.

First published 1982 by
THE MACMILLAN PRESS LTD
London and Basingstoke
Companies and representatives
throughout the world

Typeset in 10/12 Press Roman by
CAMBRIAN TYPESETTERS
Farnborough, Hants

Printed in Hong Kong

ISBN 0 333 28213 2
ISBN 0 333 28214 0 pbk

The paperback edition of the book is sold subject to the condition that it shall not, by way of trade or otherwise, be lent, resold, hired out, or otherwise circulated without the publisher's prior consent in any form of binding or cover other than that in which it is published and without a similar condition including this condition being imposed on the subsequent purchaser.

Contents

Preface	vii
Abbreviations	ix
List of Algorithms	x
1. Introduction	1
2. Some Basic Concepts	13
2.1 Graphs, Paths and Chains	13
2.2 Matrices and Computer Representation of Graphs	21
2.3 Spanning Trees	29
2.4 Multi-stage Problems and Search Trees	35
Exercises	43
3. Branch-and-bound Methods	47
3.1 Concepts of B & B	47
3.2 Integer Linear Programming	57
Exercises	65
4. Shortest Route Problems	69
4.1 Shortest Path between Two Points	70
4.2 The Shortest Path Problem: General Case	78
4.3 Other Shortest Path Problems	84
Exercises	91
5. Location Problems	94
5.1 Single Facility Problems	95
5.2 Ordinary Location Problems	101
5.3 Location of Emergency Facilities	109
Exercises	118
6. Project Networks	121
6.1 Critical Path Methods	121
6.2 Alternative Approaches	129

6.3 Resource Allocation	133
Exercises	144

7. The Travelling Salesman and Chinese Postman Problems — 148

7.1 Reduction-based Methods for Solving TSP	148
7.2 Other Approaches to TSP	157
7.3 The Chinese Postman Problem and Matching	163
Exercises	169

8. Distribution Problems — 173

8.1 Single-depot Vehicle Routing Problems: TSP and Savings Based Methods	174
8.2 Angular Approaches to Vehicle Routing	185
8.3 Multi-depot Distribution Problems	190
Exercises	193

9. Flows in Networks: Basic Model — 197

9.1 Complete Flows and Maximal Flows	197
9.2 Algorithms for Finding Maximal Flows	206
Exercises	212

10. Network Flow: Extensions — 214

10.1 Various Extensions	214
10.2 Minimal Cost Flows	288
10.3 The Simplex Method Applied to Network Problems	240
Exercises	252

11. Heuristic Methods — 255

11.1 Improvement Methods	256
11.2 Constructive Heuristic Methods	261
11.3 Problem Reduction: AND–OR Graphs	271
Exercises	279

Appendix: Computational Complexity — 282

References — 288

Index — 298

Preface

Graphs and networks are used as models in many fields of study as shown by the recent collection of articles edited by Wilson and Beineke (1979). This book is concerned with the application of graphs and networks to some of the problems that arise in industry, local government, transport and other areas. The term 'operations research' (to be interpreted in a wide sense) is adopted in the title to indicate the area from which topics are selected.

Books have been published for a number of years on applied graph theory. The field has grown to the extent that books devoted to applications in particular subjects are fully justified. Indeed I found that, within the size limitation imposed, I had to be selective as to the OR topics to be covered. The final choice is a personal one and reflects my interests and knowledge; no doubt others would have come up with a different selection.

The text grew out of a course that has been given for several years to final-year undergraduates in operations research and computing. It is designed primarily for courses in OR, but it will also be of interest to those on computing and transport studies courses as well. It is also hoped that the practising operations researcher will find it of value, and effort has been devoted to providing realistic examples (though necessarily scaled down in size).

The background mathematical knowledge required has purposely been kept to a minimum in order to make the material accessible to a wider readership. The principal prerequisites are a working knowledge of set theoretic notation and matrices, and exposure to a first course on linear programming. A prior knowledge of duality concepts would be helpful but is not necessary, the required dual results being developed as they are needed (section 10.3 is an exception to this general rule).

It has been convenient to adopt a variety of abbreviations. Some are general, some are names of algorithms (usually derived from the originators' names) and some (such as SCP for the set covering problem) are names of standard problems. Lists of these abbreviations are given on p. ix. The reader should consult these lists if ever there is any doubt as to the meaning of a particular abbreviation.

During the preparation of this book I have greatly benefited by the advice and comments of colleagues, and it is a pleasure to acknowledge this indebtedness. I would specifically like to mention Donald Davison, Steve Filbin, Chris Pursglove, Graham Rand, Grahame Settle and Derek Yates. I would also like to express my gratitude to Iain Buchanan who, as referee, made many helpful suggestions leading to the improvement of the book. Thanks are also due to Mrs Betty Jones and Mrs Pam Billingsly for typing the manuscript and to David Sherratt and Ken Chan for assistance with the GATE text editing system; they were always willing to help. Finally I would like to thank my wife Pamela for her help and support — it is to her and to my parents that the book as a whole is dedicated.

ACKNOWLEDGEMENTS

The diagrams of figure 1.2 are taken, by permission, from the *Operational Research Quarterly* from the articles indicated in the caption. Figure 1.5 is reprinted, by permission, from M. Folie and J. Tiffin, Solution of a multi-product manufacturing and distribution problem, *Management Science*, **23**, 3 (Nov, 1976), © 1976 The Institute of Management Sciences. Several exercises are (usually modified versions of) questions from University of Liverpool examination papers; these are indicated by the letters LU followed by the year in which they were set. Finally, example 6.1 is due to Frank Wharton.

Permission to use the above material is gratefully acknowledged.

Abbreviations

GENERAL

B & B	Branch-and-bound
BF	Breadth-first
CPM	Critical path method
DF	Depth-first
DK(f)	Dinic–Karzanov network
ES (EF)	Earliest start (finish) of an activity
ILP	Integer linear programming problem
LB	Lower bound
L(G)	Arc-to-vertex dual of G
LP	Linear programming problem
LS (LF)	Latest start (finish) of an activity
MILP	Mixed integer linear programming problem
MST	Minimal spanning tree
PD	Project duration
PERT	Project evaluation and review technique
SD-tree	Shortest distance tree

PROBLEMS

AP	(Linear) assignment problem
ATSP	Asymmetric travelling salesman problem
1-CD	1-centdian problem
p-CP	p-centre problem
1-CPP	1-centre problem in the plane
CPP	Chinese postman problem
LSCP	Location set covering problem
MSCP	Maximal set covering problem
p-MP	p-median problem
1-MPP	1-median problem in the plane
OTSP (s,f)	Open travelling salesman problem (from s to f)
SCP	Set covering problem
SLP	Simple location problem
TRP	Transshipment problem

TSP	Travelling salesman problem
VRP	Vehicle routing problem

ALGORITHMS

AB	To solve AP using alternating bases
AT	To find alternating trees
BF	Breadth-first search subroutine for algorithm (FF)
BKE	To solve SLP (Bilde–Krarup–Erlenkotter)
BP (FP)	Backward pass (forward pass) for CPM
CRASH	For reducing project durations
CW	Savings method for VRP (Clarke–Wright)
D	To find a shortest path (Dijkstra)
Fd (Fd′)	To find a shortest path (Ford)
FF	To find a maximal flow (Ford–Fulkerson)
Fk	To order vertices of a project network (Fulkerson)
Fl	To find all shortest paths (Floyd)
FW	For non-linear minimum cost flow problems (Frank–Wolfe)
G	To find 1-median of a tree (Goldman)
GNR	To find a p-centre (Garfinkel–Neebe–Rao)
HM	Hungarian method to solve AP
K	To find an MST (Kruskal)
L	To level resource profiles
N	To find a shortest path (Nemhauser)
OOK	Out-of-kilter algorithm for least cost flow
P	To find an MST (Prim)
PT	To form a precedence tree
RA	For resource constrained problem
SWEEP	The Gillett–Miller angular approach to VRP
Y	To find the K shortest paths (Yen)

NOTATION

ϕ	The empty set
2^X	Set of all subsets of X
$\{x \mid P\}$	Set of all elements x which satisfy condition P
$\lceil \ \rceil (\lfloor \ \rfloor)$	Smallest integer not less than (largest integer not greater than)
$\lvert S \rvert$	Number of elements in set S

1 Introduction

What is a graph? Consider first the commonly used concept defined in terms of Cartesian coordinates; figure 1.1a shows a graph of a real valued function f of a real continuous variable x where

$$y = f(x) = x^3 - 6x^2 + 9x + 16 \qquad -1\tfrac{1}{4} \leqslant x \leqslant 4\tfrac{3}{4}$$

Figure 1.1b also shows a graph, this time of

$$y = g(x) = 2x \bmod (5) \qquad 0 \leqslant x \leqslant 4$$

In both cases x can take on a (continuous) infinity of values and the graph consists of continuous lines containing an infinite number of points.

Consider again the function g, this time with x restricted to the set $\{0, 1, 2, 3, 4\}$. The graph is now given by the five points marked on figure 1.1c. Of course, since x is restricted to such a small range, all values of $g(x)$ can be listed explicitly

$$g(0) = 0, \quad g(1) = 2, \quad g(2) = 4, \quad g(3) = 1, \quad g(4) = 3$$

and the operation of g can be illustrated as in figures 1.1d and e. These provide alternative ways of showing the graph of g on $\{0, 1, 2, 3, 4\}$ and such representations are feasible when x can take on discrete values only. Such representations, which are akin to the popular concept of a network, will be those of relevance in this book.

The central concept of a graph, as understood in graph theory and throughout this book, is that it is a set of entities called *vertices*, which are interrelated via certain correspondences called *links*. A graph is often thought of as a set of points (representing vertices) with interconnecting lines (representing links) and is frequently pictured in this way. Links may be directed (with directions shown by arrows), in which case they are called *arcs*, or they may be undirected and called *edges*. Graphs are directed (undirected) if their only links are arcs (edges). Graphs can also be pictured in other ways, as shown by the examples in figure 1.2, all of which are taken from an OR journal (*Operational Research Quarterly*).

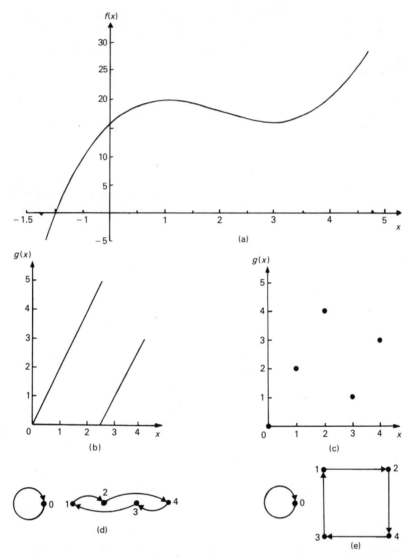

Figure 1.1 (a) Graph of $f(x) = x^3 - 6x^2 + 9x + 16$ for $-1¼ \leqslant x \leqslant 4¾$; (b) graph of $g(x) = 2x \bmod(5)$ for $0 \leqslant x \leqslant 4$; (c), (d) and (e) give three representations of $g(x)$ restricted to $x \in \{0,1,2,3,4\}$

Example (a) is an idealised diagram of a computer with a network of connections via multiplexers to local input stations. The graph is undirected and no closed curves are formed by the edges; connected graphs with this property are called *trees* (section 2.3). Vertices are represented in three ways — by dots, open circles and a rectangle.

Example (b) gives a flow chart of a queueing theory model of the internal transport system of a steelworks. The graph is directed, the arrows giving the directions of the arcs. Vertices are represented by rectangular and diamond-shaped boxes and with text written inside the boxes providing *labels*, or names, for the vertices.

Example (c) is an adaptation of a decision tree put forward in relation to a decision as to whether or not to expand rev-counter production capacity by installing extra equipment or by putting employees on overtime. Vertices are represented by large dots and rectangles, and are labelled. Extra information is given alongside the vertices.

Example (d) shows the feedback loops of a systems dynamics model of the shipping industry. The graph is directed and the arcs form many closed curves, or circuits. Vertices are identified merely by having incident arcs and by their labels (the items of text). Arcs are shown in this example by curved rather than straight lines.

Thus it is seen that pictures of graphs can take a variety of forms. Vertices may be represented in a variety of ways and may or may not be labelled. The disposition of the vertices relative to each other is not a property of graph theory and different placings, although leading to different pictures, represent the same graph. Arcs and edges may be represented by straight or curved lines and may or may not be labelled.

Six problems will now be posed to illustrate the wide range of situations to which graph theory is applicable.

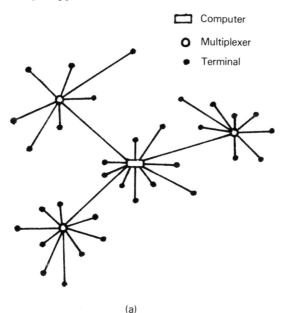

(a)

Figure 1.2 Examples of graphs from: (a) Drinkwater (1977); (b) Corkindale (1975); (c) Moore and Thomas (1973) − modified; (d) Taylor (1976)

4 Graph Theory in Operations Research

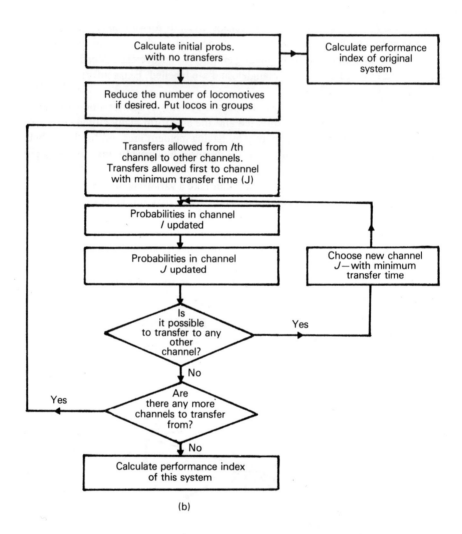

(b)

Figure 1.2 *continued*

Introduction

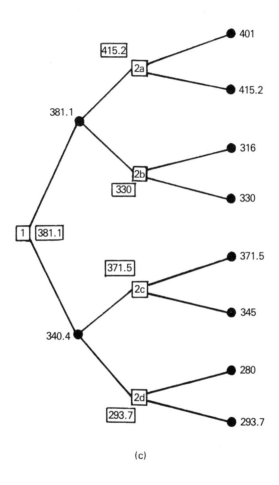

(c)

Figure 1.2 *continued*

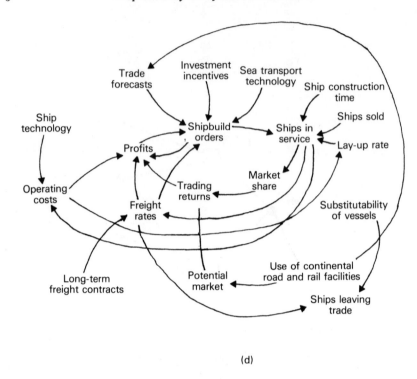

(d)

Figure 1.2 *continued*

(1) Layout Planning and Planarity (Seppänen and Moore, 1970)
An architect was asked to design a special-purpose one-storey building incorporating 5 rooms $A, B, \ldots, E$ of specified sizes etc., in such a way that each room has an outside wall and

    A is next to B, C and D
    B is next to A and C
    C is next to B, A and D
    D is next to A, C and E
    E is next to D

The architect produces the plan shown in figure 1.3a. At this stage the client notices that B is relatively far away from D, and thinking this to be a disadvantage asks whether the plan can be rearranged so that B is next to D also. The architect thinks that this is not possible but is not sure. By showing that the so-called dual graph (section 6.2) of figure 1.3b is non-planar (section 2.1), the impossibility is established.

    Similar problems arise when assigning work areas on a shop floor to satisfy specified adjacencies (Seppänen and Moore 1970, Francis and White 1974).

(2) Production Planning

A manufacturing company is asked to supply d_i units of a bulky item at the end of each of the next 6 months, that is at times $i = 1, \ldots, 6$. The company has limited storage capacity for this item which has the effect of restricting the stock on hand, s_i, at the start of each month to at most 5 units and a stock holding cost of 1 is incurred per unit per month stocked. The initial stock is 3 units and it is decided to run the stock down to zero after the present contract has been honoured. Because of variations in the cost of labour, raw materials, etc., it is estimated that the cost p_i of producing a unit in month i is given by

Month	1	2	3	4	5	6
d_i	1	1	0	3	3	4
p_i	11	13	13	12	14	13

If it is possible to produce at most 2 units in any month what should the production policy be?

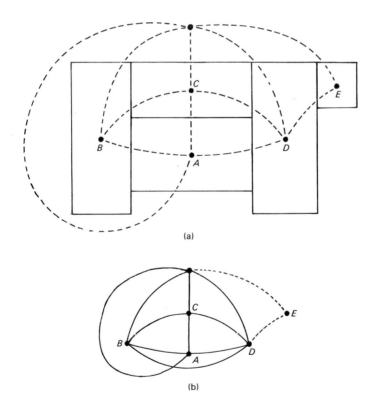

Figure 1.3

Figure 1.4 shows the possible 'states' s_i that the stock on hand can take, while satisfying all the constraints, together with the possible transitions between states. The number alongside each arc is the sum of the production and holding costs; for example, for the transition from $s_5 = 3$ to $s_6 = 2$ of month 5, 3 units are supplied to the customer and 2 units are produced, giving a total cost of

$$(s_5 \times 1) + (2 \times p_5) = 3 \times 1 + 2 \times 14 = 31$$

The problem then resolves to that of finding the 'shortest' (that is least cost) path from the state at the start of month 1 to the state at the start of month 7. The shortest paths to each state are given in figure 1.4 by the heavy lines, and it is seen that the optimal policy is to produce 2, 0, 2, 2, 1, 2 units in months 1 to 6 respectively with a total cost of 25+4+29+29+18+28=133. Algorithms for shortest paths between two specified points are given in section 4.1, and other shortest path problems are dealt with in sections 4.2 and 4.3.

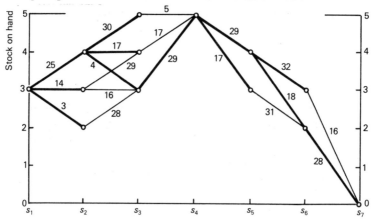

Figure 1.4 Possible transitions between states for the production planning problem

(3) Location of Bank Accounts (Cornuejols *et al.*, 1977)
In a large country, such as the United States, the number of days, α_{ij}, required to clear a cheque issued at a bank in city i and cashed at a bank in city j will vary with i and j. To maximise available funds a company that pays bills to numerous clients in various locations may find it advantageous to maintain accounts in several strategically located places (*Business Week*, 1974). If

f_i is the fixed cost of maintaining an account at city i
k_j is the volume (in dollars) of cheques cashed at city j
$c_{ij} = k_j \alpha_{ij}$
$y_i = \begin{cases} 1 \text{ if there is an account at city } i \\ 0 \text{ otherwise} \end{cases}$

Introduction

$$x_{ij} = \begin{cases} 1 \text{ if cheques to be cashed at city } j \text{ are issued at } i \\ 0 \text{ otherwise} \end{cases}$$

and p is the maximum number of accounts that can be maintained, then the problem of maximising available funds becomes

$$\text{maximise} \quad \sum_{i,j} c_{ij} x_{ij} - \sum_i f_i y_i$$

subject to

$$\sum_i x_{ij} = 1$$

$$1 \leqslant \sum_i y_i \leqslant p$$

$$y_i - x_{ij} \geqslant 0$$

$$x_{ij}, y_i \in \{0, 1\}$$

Let G be a graph with vertices $1, 2, \ldots, n$, corresponding to cities, and arcs $ij, i \neq j = 1, 2, \ldots, n$, with length c_{ij}. Then a feasible solution corresponds to any set of $m \geqslant n - p$ arcs with not more than one arc from any vertex and none from a vertex with an incoming arc. Accounts are opened at each vertex *without* an outgoing arc, and the arcs give assignments of clients to accounts.

Essentially the same problem arises in the location of various facilities, for example warehouses and post offices (sections 5.1 and 2).

(4) Distribution

A supplier of animal foodstuffs has to supply a number of regular customers, whose orders vary in size from week to week, from a central depot. Orders for delivery on a particular day are to be scheduled so as to minimise costs subject to vehicle capacities, maximum journey times, etc. How should the vehicles be routed and scheduled? Taking a suitable road map as model graph, a set of closed paths, corresponding to vehicle journeys, must be found. Each path must 'start and finish' at the depot vertex and satisfy capacity constraints, etc. Vehicle routing is discussed in chapter 8.

Higher-level decisions relate to the way in which statistical variation in demand is taken into account, in particular what should the fleet size be in the light of hiring costs resulting from unusually high demand?

(5) Multi-product Manufacturing and Distribution Problem (*Folie and Tiffin, 1976*)

A manufacturer produces one type of snack food under two brand names in a variety of different weight packages. These are produced at 8 factories which have several packing machines capable of packaging a limited range of products. It is the policy of the company to offer the full range of products to each consumer even though the system in each of the marketing regions does not always have the capacity to produce the full product range. Figure 1.5, which is taken from Folie and Tiffin (1976), relates to the specific case of potato crisps produced by Associated Products Distribution Pty Ltd of Australia. The packing machine links specify which product can be manufactured at each of the factories.

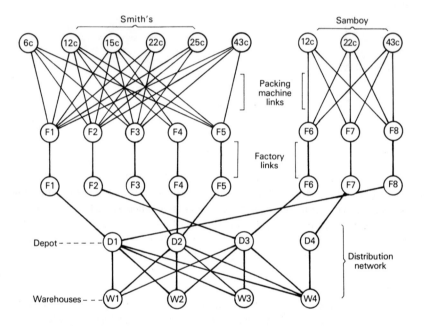

Figure 1.5 Multi-product manufacturing and distribution problem

Each packing size capability at each factory is limited by engineering factors which impose a maximum capacity on each of the packing machine links while cooking facilities limit total factory outputs. Different types of cooking and packing machinery at different factories result in varying production costs, and different package weights lead to product-dependent transportation costs. Thus the direct cost of a product at a regional warehouse is a function of the machine it was packaged on, the factory it was produced at and the route over which it was transported.

The problem is to determine flows from the 9 product vertices to the 4 regional warehouse vertices at an over-all minimum cost. This is a multi-commodity network flow problem (chapters 9 and 10).

Introduction

(6) Hospital Layout (Elshafei, 1977)

Patients attending the out-patients department of a large hospital have to move among clinics within the department. The department becomes overcrowded and the locations of the clinics relative to each other have been criticised as requiring too much travelling by patients, causing bottlenecks and serious delays. The yearly flow, f_{ij}, between clinics i and j and the distance $d(i,j)$ between i and j are known, as are the possible locations for clinics. How should the clinics be located?

We assume that each location can accommodate just one clinic and that the number, n, of available locations is equal to the number of clinics m. (If $n > m$ then $n - m$ dummy clinics can be introduced, and if $m > n$ the problem is not feasible.) The over-all movement among clinics can be minimised by solving

$$\text{minimise} \quad \varphi = \sum_{a,b} \sum_{p,q} f_{ab} d(p,q) y_{ap} y_{bq}$$

subject to

$$\sum_i y_{ij} = \sum_j y_{ij} = 1$$

$$y_{ij} = \begin{cases} 1 & \text{if clinic } i \text{ is located at } j \\ 0 & \text{otherwise} \end{cases}$$

This is an example of the classic (Koopmans–Beckman) quadratic assignment problem which has proved to be very difficult to solve and exact methods are limited to problems with n no larger than about 15 or 16. For the hospital considered by Elshafei there were 17 clinics in the out-patients department and he adopted a heuristic method for solving the problem. Certainly for larger problems exact methods are impracticable and heuristic approaches must be adopted. The question arises as to how far the approximate solution obtained may be from being optimal and this general topic is touched on in chapter 11.

The above, and many other problems will be treated in the following chapters. For a general background to operations research the reader may consult Wagner (1969) or Taha (1976).

The present book is unusual in being directed specifically towards OR applications, and encompasses a wide variety of topics for which the reader would otherwise have to consult a number of books and research papers. Accounts of parts of the material of this book, though not always from the same standpoint, are contained in Eilon *et al.* (1971), Scott (1971), Lawler (1976), Handler and Mirchandani (1979) and Mandl (1979). Also for further reading in the general area covered, Christofides (1975) and Minieka (1978) are particularly recommended. For those with some background in computing, Aho *et al.* (1974)

and Reingold *et al.* (1977) provide much useful material on implementing algorithms. Finally, Karp (1975) lists some problems which are computationally complex (see appendix) and references on heuristic approximation of computationally complex problems are given in Garey and Johnson (1976).

2 Some Basic Concepts

Graph theory suffers from a mass of definitions. The situation is made worse by different authors using different terminologies often resulting in a given word being used to convey more than one concept. It is with the basic definitions that the situation is worst and in order to provide a sound framework, formal definitions of some of the more fundamental concepts will be given in the following sections. The terminology to be used will generally follow that of Berge (1962) and Kaufmann (1967).

2.1 GRAPHS, PATHS AND CHAINS

The concept of a graph was introduced informally in chapter 1. The term will now be defined more precisely.

Definition 2.1
A *directed graph* consists of a pair of sets (V, A) where

(1) V is non-empty;
(2) elements of A are directed pairs of, not necessarily distinct, elements of V.

Elements of V and A will be called *vertices* and *arcs* respectively. $(x, y) \in A$, or xy as it is more usually written, is an arc *from x to y*. Associated with $G = (V, A)$ there is the *successor* mapping $\Gamma : V \longrightarrow V$ where the image of x under Γ, denoted by Γx, is defined by

$$\Gamma x = \{y \mid xy \in A\}$$

If $y \in \Gamma x$ then y is said to be an *(immediate) successor* of x and x an *(immediate) predecessor* of y. An alternative definition of a directed graph can now be given.

Definition 2.1a
A *directed graph* consists of a pair (V, Γ) where

(1) V is non-empty;
(2) Γ is a mapping from V to itself $(\Gamma : V \longrightarrow V)$.

In any given situation the definition which is the more convenient will be used. Related to graph $G = (V, \Gamma)$ is its *reverse* $G^{-1} = (V, \Gamma^{-1})$ of G, where $x \in \Gamma^{-1} y$ if and only if $y \in \Gamma x$. If G and its reverse coincide then G is *symmetric*

14   *Graph Theory in Operations Research*

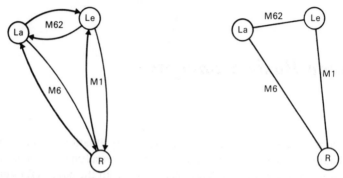

Figure 2.1 Directed and undirected graph representations of a small part of the British motorway system.
(La ≡ Lancashire, Le ≡ Leeds, R ≡ Rugby)

(see figure 2.1), and in this case it is often convenient to use the simpler corresponding undirected graph.

Definition 2.2
An *undirected graph* is a pair of sets (V, E) where

(1) V is non-empty;
(2)   elements of E, called *edges*, are undirected pairs of not necessarily distinct elements of V.

Sometimes *mixed* graphs, in which some links are directed and some undirected, are useful, an example being a city centre street map with some of the streets being one-way. From now on the adjectives 'directed' and 'undirected' will usually be omitted, the type of graph intended being clear from the context. Also, all graphs considered in this book will be *finite* in the sense that the vertex set V contains a finite number of elements.

Planarity

Diagrams such as those of figure 2.1 are merely *pictures* of graphs in a plane. It must be stressed that, for a particular graph, there is an infinity of different pictures, and two pictures of a graph can appear quite different (figure 2.2). Since pictures are used as an aid for human understanding some will clearly be better than others. Two desirable aims are that

(1)   there be few 'crossings' of arcs or edges; and
(2)   arcs or edges be represented by straight lines or at least lines which are not too contorted,

and so picture (a) is clearly the best of the three shown in figure 2.2.

Regarding (2) it has been shown by Fary in 1948 that a graph G without *loops* (edges of the form xx) can be pictured with all edges represented by straight lines. (A proof can be found in Marshall, 1971.)

Some Basic Concepts   15

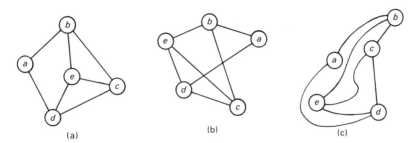

Figure 2.2 Three pictures of the same graph

Definition 2.3
A graph is *planar* if, from the infinity of possible pictures, it has a picture without crossings.

Example 2.1 (Utilities graph or $K_{3,3}$) The picture in figure 2.3a shows the graph, often called the utilities graph, or $K_{3,3}$, of connections between 3 utilities (**E**lectricity, **G**as and **W**ater) and 3 factories (F_1, F_2, F_3). Is there a picture of this graph without crossings?

Solution Suppose the vertices have been fixed in the plane, for example as in figure 2.3b, and that edges $EF_1, F_1 G, GF_2, F_2 W, WF_3$, and $F_3 E$ have been inserted without incurring any crossings. This set of edges forms a closed curve dividing the plane into two parts, an 'inside' and an 'outside'. In order that no crossings result EF_2 must lie entirely inside and GF_3 entirely outside $EF_1 GF_2 WF_3 E$ or vice versa. Without loss of generality we assume the former, and the situation shown in figure 2.3b is reached with F_1 inside and W outside the closed curve $EF_2 GF_3 E$. Consequently there is no way of inserting WF_1 without a crossing resulting. Since no assumption was made about the disposition of the vertices it follows that this graph is non-planar.

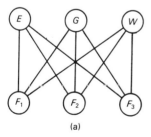
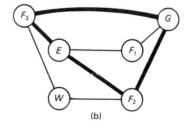

Figure 2.3 (a) The utilities graph $K_{3,3}$; (b) the remaining edge $F_1 W$ cannot be inserted without creating a crossing

Definition 2.4
The *complete (undirected)* graph K_n is the graph on n vertices which contains edge xy for every pair of vertices x and y. The *complete directed* graph on n vertices is the symmetric directed graph corresponding to K_n.

It is easily shown (exercise 2.1) that K_5 is non-planar. Moreover it was established by Kuratowski (1930) that K_5 and $K_{3,3}$ characterise *all* non-planar graphs. In order to state the result formally, first consider a vertex x of graph G which has only two incident edges ux and vx. The operation of removing x, ux and vx and adding uv, if not already present in G, will be termed a *contraction*. If G' is the graph resulting from this contraction then clearly G is planar if and only if G' is. The formal statement of Kuratowski's result is as follows.

Theorem 2.1 (Kuratowski)
A graph G is non-planar if and only if there is a graph G^*, obtained by removing edges from G, which can be reduced by a sequence of contractions to K_5 or $K_{3,3}$.

The proof of sufficiency is clear. The proof of necessity, which is more difficult, will not be given; it can be found in, for example, Marshall (1971).

Example 2.2 Use Kuratowski's theorem to show that the graph of figure 2.4 is non-planar.

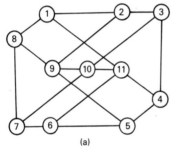 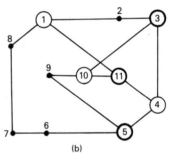

Figure 2.4 The graph of (a) is non-planar as the subgraph shown in (b) can be contracted to $K_{3,3}$

Solution Since only three vertices 9, 10 and 11 have degree in excess of 3, K_5 cannot be contained in the graph. However, if edges 2–9, 8–9, 6–11 and 7–10 are removed then the vertices 2, 6, 7, 8 and 9 each have degree 2. Performing the corresponding contractions leads to the graph $K_{3,3}$ with '1, 4 and 10 representing utilities and 3, 5 and 11 representing factories' or vice versa.

Theorem 2.1, although of much theoretical interest, does not provide an efficient means of testing for planarity. Simple but relatively weak tests are given later as theorems 2.3 and 5. Constructive approaches have proved to be more profitable. With these an attempt is made at finding a representation without crossings in such a way that if no such representation is found then the graph is non-planar; otherwise it is of course planar. Hopcroft and Tarjan (1974) produced an $O(n)$ (see appendix) constructive algorithm (see also Deo, 1976)!

Paths and Chains

Graphs in the form of road maps are usually used for finding routes between locations. Correspondingly it is convenient in graph theory to define two types of route, namely paths and chains.

Definition 2.5
A *path* from x_0 to x_s in a directed graph (where x_0 and x_s need not be distinct) is a non-null sequence of arcs $x_0 x_1, x_1 x_2, \ldots, x_{s-1} x_s$ and will be denoted by $x_0 x_1 \ldots x_s$ (or sometimes $x_0 - x_1 \ldots - x_s$).

Definition 2.6
A *chain* between x_0 and x_s in an undirected graph (where x_0 and x_s need not be distinct) is a non-null sequence $x_0 x_1, x_1 x_2, \ldots, x_{s-1} x_s$ and will be denoted by $x_0 x_1 \ldots x_s$ (or sometimes by $x_0 - x_1 \ldots - x_s$).

The term *chain* can also be used when referring to directed graphs to denote $x_0 x_1 x_2 \ldots x_s$ where *either* $x_i x_{i+1}$ *or* $x_{i+1} x_i$ is an arc (that is, directions are effectively ignored).

Definition 2.7
A *circuit* is a closed path (that is, its endpoints coincide), and a *cycle* is a closed chain. A *loop* is a circuit (cycle) containing only one arc (edge).

Definition 2.8
A path (chain) $x_0 x_1 x_2 \ldots x_s$ is *elementary* if no two of its vertices, except possibly x_0 and x_s, coincide. That is $x_i \neq x_j$ if $i \neq j$ and $\{i, j\} \neq \{0, s\}$.

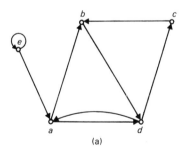
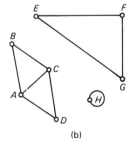

Figure 2.5 (a) A connected directed graph; (b) A non-connected undirected graph

Referring to figure 2.5, *ad, eab* and *abdcbd* are examples of paths the first two being elementary, and *AB* and *ACDA* are examples of elementary chains; *cbdc* and *cbdcbdc* are examples of circuits the first being elementary, and *bada* and *EFGE* are examples of cycles, the second being elementary. Examples of loops are *ee* and *HH*.

Various types of connectivity can be defined in terms of paths and chains, but for the present the following will suffice.

Definition 2.9
A graph G, whether directed or not, is *connected* if there is a chain between every pair of vertices of G. If G is not connected then it can be split into *components* each of which is a maximal connected graph contained in G.

Referring again to figure 2.5, the graph (a) is connected but (b) is not and consists of three components.

Generally, in a picture of a graph without crossings, there are 'empty' regions bounded by edges; such regions will be termed *faces*.

Theorem 2.2 (Euler)
For any (picture of a) connected graph without crossings, $f - q + n = 1$ where f, q and n are respectively the numbers of faces, edges and vertices.

Proof This result is readily proved by considering the effect of removing a periphery vertex (degree = 1) and all incident edges and using mathematical induction.

Theorem 2.3
For a graph G on $n \geqslant 3$ vertices to be planar $q \leqslant 3n - 6$ must hold.

Proof Consider a picture of G without crossings, and term the infinite region outside the picture the *outside*. Each face and the outside is bounded by at least three edges. Hence there are at least $3/2 (f + 1)$ edges in all, the factor $3/2$ being introduced as each edge is shared between two faces or between a face and the outside. This relationship together with theorem 2.2 gives $q = f + n - 1 \leqslant (2/3 q - 1) + n - 1$ which simplifies to $q \leqslant 3n - 6$. □

Definition 2.10
A graph G is *bipartite* if its set of vertices can be partitioned into two non-empty subsets X and Y so that there is no edge (arc) uv of G with u and v both in X or both in Y.

If all possible edges (arcs) between X and Y belong to G then G is called the *complete* (undirected or directed) bipartite graph and will be denoted by $K_{m,n}$ where m and n are the number of vertices in X and Y respectively. Undirected $K_{3,3}$ is just the utilities graph considered earlier.

Theorem 2.4
All cycles (circuits) of a bipartite graph contain an even number of edges (arcs).

Proof Suppose $u_1 u_2 u_3 \ldots u_s u_1$ is a cycle with an odd number $s = 2t + 1$ of edges. We assume without loss of generality that $u_1 \in X$ then by definition $u_2 \in Y, u_3 \in X, \ldots, u_s \in X$ and $u_1 \in Y$. But u_1 cannot belong to both X and Y so the premise that a cycle with an odd number of edges exists must be false. □

Corollary A cycle in an undirected bipartite graph must have at least four edges.

Theorem 2.5
Let G be any graph, on $n \geqslant 3$ vertices, whose cycles all have at least four edges. For G to be planar $q \leqslant 2n - 4$ must hold.

Proof The proof is similar to that of theorem 2.3 and is left as an exercise.

Corollary For a bipartite undirected graph G on $n \geqslant 3$ vertices to be planar, $q \leqslant 2n - 4$ must hold.

Example 2.3 Use theorem 2.5 to show that the graph of figure 2.4 is non-planar.

Solution It can be checked that $q=18$ and $n=11$ giving $q=2n-4$ and so the test of theorem 2.5 does not establish the non-planarity of the graph. However, removing 5–6 and contracting 4–5 to nothing (and coalescing its end vertices) results in a graph which again has cycles each with at least four edges. Now $q > 2n - 4$ and so the modified graph cannot be planar. It follows that the graph of figure 2.4 cannot be planar.

Regarding circuits and cycles in graphs we note that waste collection, mail delivery, railway inspection, etc., involve starting from some fixed point, traversing each of a set of 'roads' and returning to the starting point. Regarding roads as edges of a graph and road intersections as vertices it is relevant to pose the following question: Is it possible to find a circuit in graph G which contains each edge once and once only?

Definition 2.11
An *Euler* path (circuit) in a directed graph G is a path (circuit) which contains each arc of G once and once only. An *Euler* chain (cycle) in an undirected graph G is a chain (cycle) which contains each edge of G once and once only.

We shall consider principally the directed case, the undirected case being very similar. The existence or otherwise of an Euler path or circuit depends critically on the number of arcs incoming to and outgoing from the vertices of the graph and it is convenient to introduce the concepts of in- and out-degrees.

Definition 2.12
If x is any vertex of a directed graph G then the *in-degree* $d_i(x)$ of x is the number of arcs incoming at x and the *out-degree* $d_o(x)$ of x is the number of arcs outgoing from x.

Theorem 2.6 (Euler)
A connected graph G possesses an Euler path from s to f, where s and f are two distinct vertices of G, if and only if

(1) $d_i(s) = d_o(s) - 1$
(2) $d_i(f) = d_o(f) + 1$
(3) $d_i(x) = d_o(x), x \neq s, f$

Proof A constructive proof of the sufficiency will be given. Consider a path π which starts from s. After leaving s there will be $d_o(s) - 1 = d_i(s)$ unused arcs both incoming and outgoing from s. When any vertex, other than f, is encountered by π, one incoming arc is used and there will be [by condition (3)] a corresponding outgoing arc available. Consequently π can always be extended until finally reaching f on the last unused arc into f.

π is either the desired Euler path or there remain vertices with unused incident arcs. In the latter case let G^1 be the graph resulting from G. G^1 will consist of one or more components $C_\alpha^1, \alpha = 1, \ldots, t$, and all vertices x will have $d_i^1(x) = d_o^1(x)$. More arcs can now be added to π. Suppose $C = C_1^1$ and π have a vertex z in common. A circuit $\sigma = zu_1 u_2 \ldots u_p z$ can be found in C as, by an argument similar to that used above, the construction of a path from z can only be stopped at z. π^1 is formed from the arcs of π up to the first time it meets z, then the arcs of σ back to vertex z, then the remaining arcs of π. Then either π^1 is an Euler path or one or more components $C_\alpha^2, \alpha = 1, \ldots, r$, with $d_i^2(x) = d_o^2(x)$ all x, remain. The procedure is repeated forming a sequence of paths $\pi, \pi^1, \pi^2, \ldots$ none of which uses an arc twice and which successively contains more and more arcs. Since G is finite the process must terminate with an Euler path: this concludes the proof of sufficiency.

The proof of necessity of the conditions (1), (2), (3) follows immediately from the observation that otherwise the above constructive procedure (which will find an Euler path if one exists) must break down. □

Corollary A connected directed graph G possesses an Euler circuit if and only if $d_i(x) = d_o(x)$ for all x in G.

Example 2.4 Consider the graph of figure 2.5a (p.17)
$d_i(e) = 1, d_o(e) = 2$
$d_i(b) = 2, d_o(b) = 1$
$d_i(x) = d_o(x)$ for $x = a, c, d$

Some Basic Concepts

Thus an Euler path from e to b must exist by theorem 2.6. Suppose that in attempting to construct such a path we have chosen *eeabdcb*. This is not an Euler path and removal of used arcs leads to a graph G^1 with three isolated vertices, e, b, c and a non-trivial component C_1^1 consisting of vertices a and d and arcs ad and da. The circuit σ is clearly *ada* and π^1 is *eeadabdcb* which is an Euler path.

Definition 2.13
The *degree* $d(x)$ of any vertex x of an undirected graph G is just the number of edges incident at x.

Theorem 2.7 (Euler)
A connected undirected graph G has an Euler chain if $d(x)$ is even for all x in G, except two distinct vertices s and f. G has a Euler cycle if $d(x)$ is even for all x in G.

Proof Similar to that of theorem 2.6.

Because of the variety of basic terminology, readers, when consulting a text on graph theory, should first acquaint themselves with the particular system that the author is using.

Terms defined above are collected together, in tabular form, below.

Directed graphs	*Undirected graphs*
vertex	vertex
arc	edge
in-degree, out-degree	degree
planar	planar
path (chain)	chain
circuit (cycle)	cycle
loop	loop
elementary	elementary
connected	connected

2.2 MATRICES AND COMPUTER REPRESENTATION OF GRAPHS

In the previous section it was noted that pictures of graphs are an aid to comprehension of graph theoretical problems. For large problems however, the use of a computer becomes essential and the question arises of how best to represent a graph for computer manipulation. One way is by the use of matrices, with perhaps the simplest of these being the adjacency matrices (defined below) and their relatives.

For a directed graph (X, Γ) on vertices numbered 1 to n, a *vertex–vertex* matrix is an $(n \times n)$ matrix, P say, with the element P_{ij} giving information concerning arc ij. Examples of P include

(1) *Symbol matrix S* $S_{ij} = ij$ if $j \in \Gamma i$

                                                    $= 0$ otherwise

(Note that 'ij' is to be read as a character string rather than as a number.)

(2) *Adjacency matrix A* $A_{ij} = 1$ if $j \in \Gamma i$

                                                    $= 0$ otherwise

(3) *Distance matrix D* $D_{ij} = $ length of ij if $j \in \Gamma i$

                                                    $= \infty$ otherwise

For an undirected graph the matrix of the corresponding symmetric directed graph can be used.

Little has been achieved by introducing such matrices unless a means of operating on them is also specified. Looking first at the symbol matrix we note that $S_{ij} = ij$ if there is an arc from i to j. It would be natural to look for a rule of 'multiplication' such that $S^2_{ij} = i\alpha j$ if there is a single path $i\alpha j$ from i to j containing just two arcs, and that S^2_{ij} is a 'sum' of all such paths if there is more than one. This can be achieved as follows.

Let $\xi(uv)$ and $\eta(vw)$ be two strings of symbols which end and start respectively with v. Then if $\circ$ denotes the multiplication symbol

$$\xi(uv) \circ 0 = 0 \circ \xi(uv) = 0$$

and $\xi(uv) \circ \eta(vw)$ is formed by concatenating $\xi(uv)$ with the string that results from $\eta(vw)$ by removing the first symbol 'v' (for example, $ubv \circ vebw = ubvebw$). This is extended to 'sums' of strings by defining

$$\sum_i \xi_i(uv) \circ \sum_j \eta_j(vw) = \sum_{i,j} \xi_i(uv) \circ \eta_j(vw)$$

The product of powers of S is then defined as

$$S^{r+t}{}_{ij} = (S^r \circ S^t)_{ij} = \sum_\alpha S^r{}_{i\alpha} \circ S^t{}_{\alpha j}$$

Example 2.5   Find S and S^2 for the graph of figure 2.6a. Also find all distinct circuits containing exactly three arcs.

Solution   The symbol matrix S is given by

$$S = \begin{bmatrix} 0 & 12 & 0 & 14 & 15 \\ 0 & 0 & 0 & 24 & 0 \\ 31 & 32 & 0 & 0 & 0 \\ 41 & 42 & 0 & 0 & 45 \\ 0 & 52 & 53 & 0 & 0 \end{bmatrix}$$

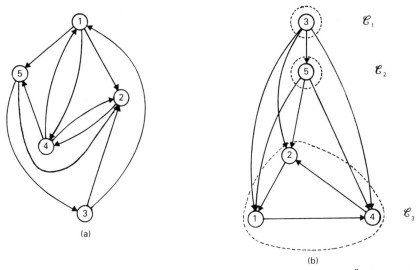

Figure 2.6 (a) The graph of example 2.5; (b) the modified graph $\tilde{G}$ obtained from the matrix (t_{ij}) (example 2.7) showing how the vertices fall into three classes

Multiplying S by itself gives

$$S^2 = \begin{bmatrix} 141 & 142+152 & 153 & 124 & 145 \\ 241 & 242 & 0 & 0 & 245 \\ 0 & 312 & 0 & 314+324 & 315 \\ 0 & 412+452 & 453 & 414+424 & 415 \\ 531 & 532 & 0 & 524 & 0 \end{bmatrix}$$

It is easily checked that this gives all paths containing just two arcs. We might expect that paths containing three arcs will be given by the elements $S^3{}_{ij}$, and, since only circuits are asked for, only the elements on the principal diagonal are shown below ...

$$S^3 = \begin{bmatrix} 1241+1531 & \cdot & \cdot & \cdot & \cdot \\ \cdot & 2412+2452 & \cdot & \cdot & \cdot \\ \cdot & \cdot & 3153 & \cdot & \cdot \\ \cdot & \cdot & \cdot & 4124+4524 & \cdot \\ \cdot & \cdot & \cdot & \cdot & 5245+5315 \end{bmatrix}$$

There are three distinct circuits with three arcs each namely 1241, 1531 and 2452 and each is given three times (starting from different vertices).

From the above example we might infer the following result.

Theorem 2.8
If S is the symbol matrix of a graph G on vertices numbered $1, 2, \ldots, n$, then $S^r{}_{ij}, r > 0$ gives all paths between vertex i and vertex j containing exactly r arcs.

Proof Follows from the definition of the law of multiplication.

Example 2.6 (Wright, 1975) A Housing Authority has a number of houses at its disposal each of which falls into one of N categories depending on location, number of bedrooms, rent, possession of garage, central heating installed, etc. Over a period of time some tenants will wish to move to alternative accommodation as their circumstances change (change of jobs, families growing up, etc.). The Authority aims to match the requirements of tenants to their houses. Although simple exchanges can satisfy some of the desired changes, longer sequences are often required. Consider the simplified case of just five types of house and the matrix

$$(C_{ij}) = \begin{bmatrix} 0 & 5 & 0 & 4 & 6 \\ 0 & 0 & 0 & 12 & 0 \\ 8 & 5 & 0 & 0 & 0 \\ 10 & 2 & 0 & 0 & 4 \\ 0 & 7 & 5 & 0 & 0 \end{bmatrix}$$

where C_{ij} is the number of tenants at present in a type i house who wish to move to a type j house. It is assumed that type 5 refers to houses not under the Authority's control and the system can be regarded as being closed. Advise on the changes of tenancy to be made.

Solution C is another example of a vertex–vertex matrix referring in this case to the graph G of figure 2.6a. Since the system is closed, for each tenant who moves from a type i to a type j house there must be one who moves from a type j to a type k house and so on until the sequence closes with a tenant moving from a type l to a type i house. Correspondingly we form the symbol matrix S of G and search for circuits in order of increasing length (in terms of number of arcs).

S has already been given in example 2.5 and it may be noted that there are just two circuits, 141 and 242 each with two arcs. Four tenants in type 1 houses 'exchange' with four tenants in type 4 houses and two tenants in type 2 houses 'exchange' with two tenants in type 4 houses. Again referring back to example 2.5, it is seen that there are three circuits, 1241, 1531 and 2452, each with three arcs. Five tenants move from each type of house in circuit 1241. Similarly five and four tenants move from each type of house in circuits 1531 and 2452 respectively.

Some Basic Concepts

Now no tenants remain who wish to make the changes type 1 to type 2, type 1 to type 4, type 4 to type 2, type 4 to type 5 and type 5 to type 3 and so the search will be continued with the corresponding arcs deleted from G. If $\tilde{S}$ is the new symbol matrix then $\tilde{S}^r$ can be obtained from $S^r, r > 0$, simply by removing terms containing 12, 14, 42, 45 or 53. Referring back to example 2.5, $\tilde{S}^2$ is given by

$$\tilde{S}^2 = \begin{bmatrix} 0 & 152 & 0 & 0 & 0 \\ 241 & 0 & 0 & 0 & 0 \\ 0 & 0 & 0 & 324 & 315 \\ 0 & 0 & 0 & 0 & 415 \\ 0 & 0 & 0 & 524 & 0 \end{bmatrix}$$

from which

$$\tilde{S}^4 = \begin{bmatrix} 15241 & 0 & 0 & 0 & 0 \\ 0 & 24152 & 0 & 0 & 0 \\ 0 & 0 & 0 & 31524 & 32415 \\ 0 & 0 & 0 & 41524 & 0 \\ 0 & 0 & 0 & 0 & 52415 \end{bmatrix}$$

This gives the single circuit 15241 and one tenant moves from each type of house, round this circuit. Since no tenant wishing to make the change type 1 to type 5, type 2 to type 4 or type 4 to type 1 remains, there are clearly no relevant circuits with five arcs, and so the procedure terminates. (Circuits with more than five arcs need never be considered of course.)

Of the 68 desired changes 58 have been realised and the numbers of desired changes not made are given by the matrix

$$\begin{bmatrix} 0 & 0 & 0 & 0 & 0 \\ 0 & 0 & 0 & 0 & 0 \\ 3 & 5 & 0 & 0 & 0 \\ 0 & 0 & 0 & 0 & 0 \\ 0 & 2 & 0 & 0 & 0 \end{bmatrix}$$

Finally, any unsatisfied changes to the 'outside' must be allowed even though not part of a circuit. In this example no desired moves to the outside (type 5) are unsatisfied and so no further changes are necessary.

The reader may have noticed that since column 3 of $\tilde{S}^2$ contains only zeros no circuit can contain vertex 3, and so column 3 and row 3 could have been omitted at this stage. The procedure adopted is heuristic in that, in general, there may be other policies leading to fewer desired changes remaining unsatisfied. Wright (1975) recommends a somewhat different strategy (exercise 4.6).

The adjacency matrix A of a graph is just the symbol matrix with each non-zero term replaced by '1'. It follows that if A^r is formed by ordinary matrix multiplication then A^r_{ij} is just the number of component strings of S^r_{ij}, that is the number of distinct paths from vertex i to vertex j containing precisely r arcs. For example, for the graph of figure 2.6a, A^2 is readily found from the corresponding matrix S^2 given in example 2.5

$$A^2 = \begin{bmatrix} 1 & 2 & 1 & 1 & 1 \\ 1 & 1 & 0 & 0 & 1 \\ 0 & 1 & 0 & 2 & 1 \\ 0 & 2 & 1 & 2 & 1 \\ 1 & 1 & 0 & 1 & 0 \end{bmatrix}$$

It is frequently useful to adopt a different rule of multiplication. Writing B in place of A and terming this new matrix the *Boolean* adjacency matrix to reflect this change, define

$$(B^{r+s})_{ij} = B^r_{i1} \times B^s_{1j} \dotplus B^r_{i2} \times B^s_{2j} \dotplus \ldots \dotplus B^r_{in} \times B^s_{nj}$$

where $\dotplus$ is the Boolean operation defined by

$$0 \dotplus 0 = 0, \qquad 0 \dotplus 1 = 1 \dotplus 0 = 1, \qquad 1 \dotplus 1 = 1$$

It can be seen that $B^r_{ij} = 0$ if $A^r_{ij} = 0$ and is 1 otherwise. If I denotes the identity matrix of size n, consider the matrix $I \dotplus B$ obtained from B by setting each element to 1 on the principal diagonal. Using the new rule of multiplication

$$(I \dotplus B)^r = I \dotplus (B \dotplus \ldots \dotplus B) \dotplus \ldots \dotplus (B^i \dotplus \ldots \dotplus B^i) \dotplus \ldots \dotplus B^r$$
$$\qquad\qquad\qquad \binom{r}{1} \text{ terms} \qquad\qquad \binom{r}{i} \text{ terms}$$

$$= I \dotplus B \dotplus B^2 \dotplus \ldots \dotplus B^i \dotplus \ldots \dotplus B^r$$

since $1 \dotplus 1 = 1$ implies $B \dotplus B = B$, etc. Interpreting I as B^0 leads to the following.

Theorem 2.9
For a graph G with vertices numbered $1, 2, \ldots, n$, $(I \dotplus B)^r_{ij}$ is 1 if there is a path from vertex i to vertex j containing no more than r arcs, and is zero otherwise.

Corollary (1) $(I \dotplus B)^r = (I \dotplus B)^{n-1} = R$ for all $r \geq n$.

(2) $R_{ij} = \begin{cases} 1 \text{ if there is a path from } i \text{ to } j \\ 0 \text{ otherwise} \end{cases}$

Some Basic Concepts

The limiting matrix R is called the *reachability* matrix of graph G.

Example 2.7 (Machine loading problem)   A company's product is sold as several different models each of which requires a given type of component in one of five closely related forms $F_1, F_2, \ldots, F_5$. Each F_i can be produced on a single specialist machine but it takes a time t_{ij} in re-setting the machine when changing over from F_i to F_j. Space limitations necessitate daily production of a batch of each $F_i, i = 1, \ldots, 5$, and for convenience the same batch order is to be followed each day. Ignoring start up and close down times, and using the following matrix of change-over times (in minutes)

$$(t_{ij}) = \begin{bmatrix} 0 & 11 & 10 & 4 & 9 \\ 9 & 0 & 11 & 8 & 10 \\ 7 & 5 & 0 & 6 & 6 \\ 6 & 4 & 9 & 0 & 12 \\ 6 & 8 & 7 & 5 & 0 \end{bmatrix}$$

find a production sequence which leads to a low (and preferably minimal) total change-over time.

Solution   Construct the complete directed graph G on five vertices $F_1, F_2, \ldots, F_5$ and assign length t_{ij} to each arc $F_i F_j$. The problem can then be regarded as that of finding a shortest path in G which visits each vertex once and once only. (Such a path is called a *Hamilton path* — see section 7.2.) There are $5! = 120$ feasible solutions and $n!$ in the general case of n vertex complete graphs. It is well known that $n!$ grows extremely rapidly as n increases (for example $11!$ is almost 40 million). Because of this it may be considered worth while accepting a sequence which might be not quite optimal in return for a reduction in effort of finding the solution.

Now if $t_{ij} > t_{ji}$ then the change-over from F_j to F_i seems more likely to appear in good (low total change-over time) sequences than the change-over from F_i to F_j. Consequently the search for a minimal length Hamilton path will be conducted in the graph $\tilde{G}$ obtained from G by deleting arc $F_i F_j$ if $t_{ij} > t_{ji}$ with ties being resolved arbitrarily. For the problem at hand $\tilde{G}$ has (Boolean) adjacency and reachability matrices

$$B = \begin{bmatrix} 0 & 0 & 0 & 1 & 0 \\ 1 & 0 & 0 & 0 & 0 \\ 1 & 1 & 0 & 1 & 1 \\ 0 & 1 & 0 & 0 & 0 \\ 1 & 1 & 0 & 1 & 0 \end{bmatrix} \quad R = \begin{bmatrix} 1 & 1 & 0 & 1 & 0 \\ 1 & 1 & 0 & 1 & 0 \\ 1 & 1 & 1 & 1 & 1 \\ 1 & 1 & 0 & 1 & 0 \\ 1 & 1 & 0 & 1 & 1 \end{bmatrix} \begin{matrix} \mathscr{C}_3 \\ \mathscr{C}_3 \\ \mathscr{C}_1 \\ \mathscr{C}_3 \\ \mathscr{C}_2 \end{matrix}$$

Clearly the rows of R fall into three categories and hence so do the vertices of $\tilde{G}$. All vertices are reachable by a path from F_3. All vertices except F_3 can be reached from F_5. All vertices except F_3 and F_5 can be reached from F_1, F_2 and F_4 (figure 2.6b, p. 23).

Clearly any Hamilton path in $\tilde{G}$ must start at F_3 then go to F_5 then to F_1, F_2 and F_4 in some order. The three feasible solutions, together with the corresponding total change-over times, are

$F_3 F_5 F_4 F_2 F_1$ cost $6 + 5 + 4 + 9 = 24$ mins

$F_3 F_5 F_1 F_4 F_2$ cost $6 + 6 + 4 + 4 = 20$ mins

$F_3 F_5 F_2 F_1 F_4$ cost $6 + 8 + 9 + 4 = 27$ mins

The production sequence chosen is $F_3 F_5 F_1 F_4 F_2$ with total change-over time of 20 minutes. In fact this can be seen to be the optimal solution since the shortest change-over times to each of $F_1, F_2, \ldots, F_5$ are 6, 4, 7, 4 and 6 respectively and the sum of the four smallest of these is 20.

A larger example along the above lines is given in Kaufmann (1967). Other, and better, methods of solving such a problem will be given in chapter 7.

Among other matrices associated with graphs are ones in which edges play a more prominent role; this is particularly true in electrical network theory. Such matrix representations are not relevant to the development presented here and will not be discussed further.

We return now to the question of computer representation of graphs. Matrices are familiar and well-understood constructs in computing which provide simple means of storing information about graphs. Also, as discussed above, their manipulation gives much information about the path structure of a graph. Indeed for graphs which are complete, or nearly so, there is much to be said for matrix operations with a vertex–vertex representation. On the other hand for *sparse* graphs, by which we shall mean graphs with the number of edges of the order of the number of vertices (or less), the use of matrices is clearly wasteful of computer storage, and also leads to unnecessary computation. To see this, consider the case of a graph on vertices numbered $1, 2, \ldots, n$, which consists simply of the one path $123 \ldots n$. It could be shown that vertex n is reachable from vertex 1 by forming the matrix $(I \dotplus B)$ and squaring $\lfloor \log_2 (n-1) \rfloor$ times to obtain the reachability matrix R. ('$\lfloor \ \rfloor$' denotes 'the largest integer not less than'.) However, the straightforward multiplication of two matrices requires $n - 1$ additions for each of n^2 elements and so requires $O(n^3)$ operations in all. Alternatively, the intuitive approach of starting at vertex 1 looking at its successors, at their successors and so on until n is encountered requires $O(n)$ operations for this example if a list of successors is kept for each vertex. (Is this intuitive approach to be preferred to the determination of R if the graph is not sparse?)

Although not always as clear cut as the above example might suggest, it is true that it is generally more efficient to store graphs in vertex-successor form if they are not complete or nearly complete. Perhaps this is not surprising as the vertex-successor representation more closely resembles the picture representation. Information stored for a particular vertex might take the form of a collection of 'records' with the structure

Name	Vertex information	P_1	P_2	. . . .	$\wedge$	S_1	S_2	. . . .	$\wedge$

$P \equiv$ predecessor, $S \equiv$ successor, $\wedge \equiv$ special symbol to denote the end of a list.

Situations in which the vertex-successor or a related representation is appropriate will often occur in later sections. Some special representations for trees are introduced in section 10.3.

2.3 SPANNING TREES

Examples of trees have been given in chapter 1. More formally, the following definition may be given.

Definition 2.14
A *tree* is a connected undirected graph without cycles.

Figure 2.7 shows all the distinct trees with six or fewer vertices; after this the number of distinct trees increases rapidly with the number of vertices. (For $n = 7, \ldots, 12$ the numbers are 11, 23, 47, 108, 235 and 551 respectively.)

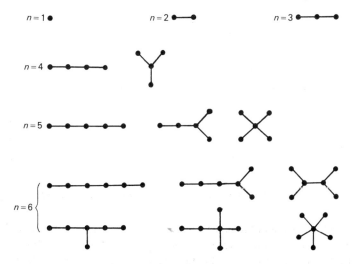

Figure 2.7 All distinct trees with six or fewer vertices

Theorem 2.10
The following statements about an undirected graph G are equivalent

(1) G is a tree;
(2) G is connected with n vertices and $n-1$ edges;
(3) G has n vertices, $n-1$ edges and no cycles;
(4) G is such that each pair of vertices is connected by a *unique* elementary chain.

Proof   The proof of this theorem is left as an exercise.

From now on (1) to (4) above will be regarded as equivalent definitions and the one used in any particular instance will be the one most appropriate to that situation.

Trees arise quite naturally when one is interested in connecting vertices in such a way that the total edge length is minimal (since the existence of a cycle would allow an edge to be removed and the condition of minimality to be violated). If no further conditions are imposed then this is the classical *Steiner problem* (Courant and Robbins, 1941) and a solution will be termed a *Steiner tree*. As an example consider the case of four vertices A, B, C and D situated at the corners of a square of side s. Then a solution is given in figure 2.8a, another solution being obtained by rotating the edges through $\pi/2$ about the centre of the square. In a Steiner tree at most three edges meet at any junction and for any junction other than an original vertex precisely three edges meet at angles of $2\pi/3$. The problem of identifying Steiner trees is in general a very difficult problem but a method for finding trees which approximate Steiner trees is given by Chang (1972).

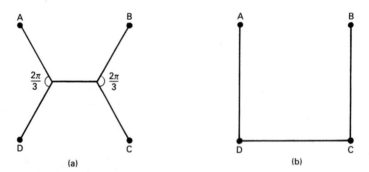

Figure 2.8 (a) Steiner tree; (b) minimal spanning tree

The problem generally remains computationally difficult even when the 'non-vertex' junctions are restricted to belonging to a finite set S (Hakimi, 1971 and Lawler, 1976). However, in the limiting case of $|S| = 0$, very efficient algorithms are available.

Some Basic Concepts

Definition 2.15
A *spanning tree* T of a *connected* graph G on vertex set V is a tree, also with vertex set V, whose only edges correspond to edges (or perhaps arcs) of G.

Edges or arcs of a graph G frequently have associated weights and it is customary in such cases to refer to G together with the set of weights as a *network*.

Definition 2.16
A *(directed) network* $N = (V, A, w)$ comprises a graph $G = (V, A)$ together with *weights* $w(xy)$ corresponding to the arcs xy of G. An *(undirected) network* $N = (V, E, w)$ comprises a graph $G = (V, E)$ together with *weights* $w(xy)$ corresponding to the edges xy of G.

In later sections the terms 'graph' and 'network' will often be used more or less interchangeably when the weight function is clearly understood.

Given any collection C of arcs (or edges) of a network N with *weight* function w then the weight $w(C)$ of C is just the sum of the weights of the elements of C

$$w(C) = \sum_{xy \in C} w(xy)$$

Definition 2.17
A *minimal* spanning tree (or MST) of a network N is a spanning tree of minimal weight.

In order to obtain an MST of a network N it may first be assumed, without loss of generality, that N is complete (that is, there is an edge between every pair of vertices of N). (If this were not the case, 'missing' edges could be inserted with weight ∞ without altering any MST.) Then the following algorithm due to Kruskal (1956) builds a series of graphs $G_1, G_2, \ldots, G_s$ with G_s being an MST.

Algorithm (K)
To find an MST of a complete network (V, E, w)

Step 1 (Setup)
    Set G_1 to be the graph with vertex set V and no edges.

Step 2 (Iteration)
    Find a minimal weight edge, xy say, of G_i.
    Set $w(xy) = \infty$
    If the addition of xy to G_i would not form a cycle then add xy to G_i to form G_{i+1} and replace i by $i + 1$.

Step 3 (Termination)

If $i < n$ return to step 2,

otherwise G_i is an MST.

Example 2.8 One of the products of a manufacturer is a piece of electrical equipment. Each item of equipment has eight terminals $T_1, T_2, \ldots, T_8$ which must have the same electrical potential. This is to be achieved by connecting the terminals with a network of wires. If the distance between terminals T_i and T_j is proportional to the Euclidean distance indicated by the grid of figure 2.9a, find a connecting network which minimises the total amount of wire used.

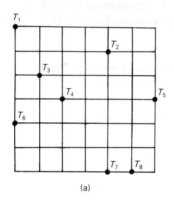 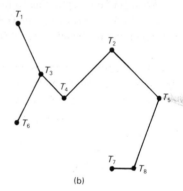

(a)   (b)

Figure 2.9

Solution Since a wire segment must connect a pair of terminals the desired network will be an MST of the complete network on vertex set $\{T_1, T_2, \ldots, T_8\}$. If E_i denotes the set of edges of the graph G_i formed by algorithm (K) then

$$\begin{aligned}
E_1 &= \phi \\
E_2 &= \{T_7 T_8\} \qquad T_7 T_8 \text{ is the shortest edge} \\
E_3 &= \{T_7 T_8, T_3 T_4\} \qquad T_3 T_4 \text{ is the next shortest edge} \\
E_4 &= \{T_7 T_8, T_3 T_4, T_1 T_3\} \\
E_5 &= \{T_7 T_8, T_3 T_4, T_1 T_3, T_3 T_6\}
\end{aligned}$$

Addition of $T_4 T_6$ would form cycle $T_4 T_6 T_3 T_4$

$$\begin{aligned}
E_6 &= \{T_7 T_8, T_3 T_4, T_1 T_3, T_3 T_6, T_2 T_4\} \\
E_7 &= \{T_7 T_8, T_3 T_4, T_1 T_3, T_3 T_6, T_2 T_4, T_2 T_5\}
\end{aligned}$$

Addition of $T_2 T_3$ would form cycle $T_2 T_3 T_4 T_2$

$$E_8 = \{T_7 T_8, T_3 T_4, T_1 T_3, T_3 T_6, T_2 T_4, T_2 T_5, T_5 T_8\}$$

There are eight vertices, seven edges and no cycles and hence by theorem 2.10 the graph G_8 with edge set E_8 is an MST (figure 2.9b). Another MST differs only in that it includes $T_4 T_6$ instead of $T_3 T_6$. (Are there any more?)

That algorithm (K) does indeed yield an MST follows from the theorem below whose proof is left as an exercise.

Theorem 2.11
Let $N = (X, E, w)$ be a complete network, and $T_1 = (V_1, E_1)$, $T_2 = (V_2, E_2)$ be MSTs over V_1 and V_2 respectively, where V_1 and V_2 are disjoint non-empty subsets of X. If ab is a minimal length edge between V_1 and V_2 then $T = (V_1 \cup V_2, E_1 \cup E_2 \cup \{ab\})$ is an MST over $V_1 \cup V_2$ if

(1) $|V_1| = 1$ (or $|V_2| = 1$), or
(2) no edge of T_1 or T_2 has weight greater than $w(pq)$ for any $pq \neq ab$ with $p \in V_1, q \in V_2$.

We now turn to a second algorithm for finding MSTs proposed by Prim (1957) and independently by Dijkstra (1959). In this a tree is 'grown' in a rather different way, three sets being used in the algorithm, namely

Q—set of vertices already in the tree
E—set of edges already in the tree
C—set of edges to be considered next

At any stage the set of edges E and the set of vertices Q together form an MST over the vertex set Q. Again without loss of generality it will be assumed that the initial network is complete.

Algorithm (P)
To find an MST of a complete network (X, E, w)

Step 1 (Setup)
  Select a vertex $a \in X$ and set $Q = \{a\}$, $E = \phi$, $C = \{ax \mid x \notin Q\}$.

Step 2 (Iteration)
  (1) Select a minimal weight edge uv in C with $u \in Q$, and transfer uv from C to E and place v in Q (that is, set $Q \leftarrow Q \cup \{v\}$, $E \leftarrow E \cup \{uv\}$, $C \leftarrow C - \{uv\}$).
  (2) For every $x \notin Q$ replace edge sx in C by the edge vx if $w(vx) < w(sx)$.

Step 3 (Termination)
  If $C \neq \phi$ then repeat step 2.
  Otherwise (Q, E) is an MST.

The validity of this algorithm follows from theorem 2.11 if V_1 is taken as Q and V_2 as $\{v\}$.

Example 2.9 Find an MST for the undirected network on vertex set $\{1,2,3,4,5\}$ whose edge weights are given by the following matrix

$$\begin{bmatrix} \infty & 7 & 11 & 3 & 12 \\ & \infty & 12 & 14 & 1 \\ & & \infty & 8 & 9 \\ & & & \infty & 4 \\ & & & & \infty \end{bmatrix}$$

Only the upper half of the matrix is given since *yx* and *xy* are the *same* edge. Each diagonal element is set to ∞ since we know that an MST cannot contain any loops.

Solution
Step
1 Let *a* be the vertex 1, then

$Q = \{1\}, \quad E = \phi, \quad C = \{12, 13, 14, 15\}$

2a 14 is a minimal weight edge in *C*

$Q \leftarrow \{1,4\}, \quad E \leftarrow \{14\}, \quad C \leftarrow \{12, 13, 15\}$

2b Since $w(43) = w(34) < w(13)$ and $w(45) < w(15)$

$C \leftarrow \{12, 43, 45\} = \{12, 34, 45\}$

2a 45 is a minimal weight edge of *C*

$Q \leftarrow \{1,4,5\}, \quad E \leftarrow \{14, 45\}, \quad C \leftarrow \{12, 34\}$

2b Since $w(52) < w(12)$

$C \leftarrow \{52, 34\} = \{25, 34\}$

2a 25 is a minimal weight edge of C

$Q \leftarrow \{1,4,5,2\}, \quad E \leftarrow \{14, 45, 25\}, \quad C \leftarrow \{34\}$

2b *C* is unchanged.
2a 34 is the only candidate as minimal weight edge of *C*

$Q \leftarrow \{1,4,5,2,3\}, \quad E \leftarrow \{14, 45, 25, 34\}, \quad C \leftarrow \phi$

2b *C* remains unchanged.
3 An MST is given by the edges of *E*.

Having presented the two principal methods for finding MSTs it is appropriate to give a short discussion of their relative efficiencies and circumstances under which each might be useful.

Suppose first of all that the edges $x_i y_i$ have previously been ordered so that

$$w(x_1 y_1) \leqslant w(x_2 y_2) \leqslant \ldots \leqslant w(x_q y_q)$$

then clearly algorithm (K) is quick to apply since it only requires $n-1$ edges [together with (a few) edges that are rejected in order to avoid cycles] to be selected from the head of the list. Also the test for cycles can be achieved quite simply and the total computational effort required to apply algorithm (K) is $O(n)$. On the other hand algorithm (P) requires $n-\alpha$ comparisons to be made in step 2b for $\alpha = 1, 2, \ldots, n-1$, and so will require a total of $O(n^2)$ operations. Thus, if the edges have been pre-sorted or are such that they can be sorted with relatively little effort or when the network is *sparse* (that is, has relatively few edges — $O(n)$ or less) — then algorithm (K) requires less computational effort.

If the edges have not been pre-sorted then the situation is rather more complex. Sorting the q edges will require $O(q \log q)$ operations (Knuth, 1973) which for $q = O(n)$ gives $O(n \log n)$ operations required for algorithm (K). However a complete sort is not necessary, and if a tree sort (Knuth, 1973) is employed then the effort is reduced somewhat. More recently, Yao (1975) has given an algorithm which is $O(n \log \log n)$.

A useful property of algorithm (K) [not shared by algorithm (P)] is that it automatically produces, as a byproduct, minimal spanning m-forests.

Definition
A *minimal spanning m-forest* of a complete undirected network $N = (X, E, w)$ is a collection of m MSTs, $T_i = (V_i, E_i)$ over $V_i, i = 1, \ldots m$, with respect to w where the sets V_i are disjoint and together cover the whole of X.

This concept has relevance in the field of cluster analysis (Gower and Ross, 1969). Two entities x and y are often regarded as belonging to the same cluster if they are sufficiently close together, say $d_{xy} < \delta$ for some δ. Suppose now that δ is fixed and that algorithm (K) is applied but with termination as soon as an edge whose length exceeds δ is selected for possible addition. The $m(\delta)$ vertex sets of the MSTs comprising the resulting minimal spanning $m(\delta)$-forest are the required clusters. Alternatively, if the number of clusters, m, is fixed then a minimal separation $\delta(m)$ yielding the m clusters can be found.

2.4 MULTI-STAGE PROBLEMS AND SEARCH TREES

The solution of many problems involves the transformation of a system from one state to another, often via well-defined intermediate states. In such cases a directed graph (X, Γ) forms a natural model, with vertices representing states of the system and arcs possible transitions between states. Typically there is a weight r_{ij} associated with each arc ij representing cost, time, distance, maximum flow, reliability, etc. (Illustrations of these interpretations of weight are given in subsequent pages.) We also suppose that the problem requires the minimisation of some function $f(\pi)$ over some set of paths $P = \{\pi_i\}$.

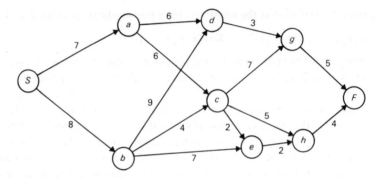

Figure 2.10 Find a shortest path from S to F. The length d_{xy} is given alongside each arc xy

Example 2.10 (Route Planning)   Mr Thompson wishes to travel from S-town to F-bridge by the shortest route possible given the information of figure 2.10.

Solution   From S, Mr Thompson can first travel to a, and thence via c or d, or to b and thence via c, d or e. These alternatives and all their possible continuations are illustrated in figure 2.11. A path from vertex S to any 'terminal' vertex labelled F corresponds to precisely one route from S to F through the network of figure 2.10, and an optimal solution is represented by the decisions 'go to b', 'go to c', 'go to e', 'go to h' and 'go to F'.

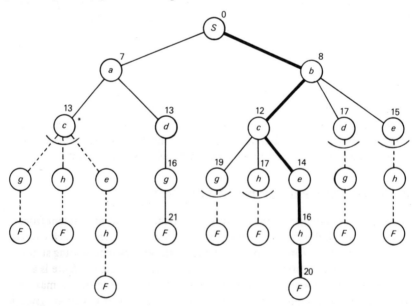

Figure 2.11 The search tree for example 2.10. Cutoffs are indicated by arcs of circles and the unique optimal solution is distinguished by the heavy lines

Some Basic Concepts

Suppose for the moment that the route is constrained to pass through *c*. Then to every route starting with the sub-route *Sac* there is a corresponding route, starting with sub-route *Sab*, which is shorter (*SbcgF* is shorter than *SacgF*, *SbchF* is shorter than *SachF* and *SbcehF* is shorter than *SacehF*) and so no route starting with *Sac* can be optimal. Since this situation is apparent as soon as the lengths of *Sac* and *Sbc* have been determined the paths of the complete enumeration tree which start with *Sac* need not be further explored and we say they have been *implicitly enumerated*, and that a *cutoff* has occurred.

The above reasoning gives an example of the general result, often called *the Principle of Optimality*, that an optimal solution (route) consists only of optimal sub-solutions (sub-routes) (Bellman, 1957). By similar arguments the tree can be further 'pruned' as indicated in the figure leaving a smaller tree, the *search tree*, which is searched for an optimal solution. That is, it was necessary actually to consider only 15 out of the 28 vertices; thus in some sense, the effort required has been reduced by about 46 per cent.

Concepts relating to search trees can be described, in a form appropriate for developments in later chapters, by means of partitions.

Definition 2.18
A *partition* of a set A is any collection $\{B_i\}$, $i = 1, 2, \ldots, t$ of subsets B_i of A such that

$$B_1 \cup B_2 \cup \ldots \cup B_t = X \quad \text{(covering property)}$$
$$B_i \cap B_j = \phi \text{ if } i \neq j \quad \text{(disjointness property)}$$

A partition is *proper* if $B = \phi$ for no i, and $t > 1$.

Let Q be a finite non-empty set. Suppose that Q is properly partitioned into subsets $Q_1, Q_2, \ldots, Q_t$, that sets Q_i are properly partitioned into subsets $Q_{i_1}, Q_{i_2}, \ldots, Q_{i_s}$ and so on. Corresponding to this partitioning a tree $T = (V, E)$ can be grown. V is formed by associating an element $t(A)$ with every subset A of Q (including Q itself) that is generated. E comprises $\{t(A)t(B_i)\}$ corresponding to partitions $\{B_i\}$ of all sets A.

Definition 2.19
Suppose Q is a finite set which contains all feasible solutions to a problem P. A *search tree* for P is any tree that can be grown by proper partitioning starting with the set Q.

Vertices of search trees will now be called *nodes* to distinguish them from vertices of other graphs.

Definition 2.20
Let $T = (V, E)$ be a search tree for problem P grown from set Q. The *root* of T

is the node $t(Q)$. A node, other than the root, is a *leaf* if it has only one incident edge. A *terminal* is a leaf y associated with a set $B \subset Q$ [$y = t(B)$] where B contains only a single feasible solution to P. A *cutoff* occurs at any leaf node which is not also a terminal. A search tree for problem P is a *complete enumeration tree* for P if all its leaves are terminals.

Example 2.11   The full tree of figure 2.11 is a search tree for the problem of example 2.10 with Q as the set of all paths from S to F. The root is the node labelled S. Let

Q_1 be the set of all paths from S to F via a
Q_2 be the set of all paths from S to F via b

Clearly $\{Q_1, Q_2\}$ is a proper partition of Q. Q_1 and Q_2 are the sets corresponding to the nodes labelled a and b. That is, $a = t(Q_1)$ and $b = t(Q_2)$. Any node labelled F is a terminal. Since the only leaves are terminals this is a complete enumeration tree.

Another example of a search tree for the same problem is obtained from the tree of figure 2.11 by ignoring edges marked by broken lines and the consequent isolated vertices. The starred node in figure 2.11, labelled c, is now a leaf but not a terminal — a cutoff has occurred.

Ways in which the size of search trees can be reduced by implicit enumeration include (discrete) dynamic programming (see below) and branch-and-bound (chapter 3). If the guarantee of optimality is relaxed, the reduction of size of search trees can be obtained by heuristic search (chapter 11).

Discrete Dynamic Programming

Many decision problems can be split into a series of *stages* $1, 2, \ldots, n$ with a decision d_i having to be made at each stage i. At stage i the *state* s_{i-1} of the system, resulting from 'earlier' decisions $d_1, \ldots, d_{i-1}$ must in general be known in order to make decision d_i effectively. We write

$$s_i = d_i s_{i-1} \qquad i = 1, \ldots, n$$

The proper definition of 'state' is crucial and if the state s_i cannot be determined by a knowledge of s_{i-1} and d_i only, then the states at stage $i-1$ are incomplete or incorrectly defined. That is, no explicit reference to prior decisions need be made, the necessary information being completely given by the state s_{i-1}. In graph theoretical terms the states are represented by vertices and the transitions $s_{i-1} \to s_i$ by arcs. Very often a stage *return* $r_i(s_{i-1}, s_i)$ can be associated with each arc $s_{i-1} s_i$ and the problem can be expressed as the minimisation (or maximisation) of a function φ of these stage returns. Some at first sight non-multi-stage problems can be put in this form by imposing an order in which the component decisions are to be made.

Example 2.12 (Knapsack problem) A company has a small factory with only a limited amount of machining capacity. It is decided that m products $X_1, X_2, \ldots, X_m$ are to be produced at this factory in the coming year. The profit per unit of X_i is p_i, each unit of X_i requires a machine time of t_i, and T units of machine time are available per day. Assuming all items produced can be sold and that it is not feasible to split production of an item over 2 days determine a product mix which yields maximum profit for the specific data set below.

i	1	2	3	4	
p_i	5	11	7	17	
t_i	2	4	3	5	$T = 8$

Solution Denoting by x_i the number of items of X_i produced, the problem can be formulated as

KP: maximise $p_1 x_1 + p_2 x_2 + \ldots + p_m x_m$ (2.1)

subject to $t_1 x_1 + t_2 x_2 + \ldots + t_m x_m \leq T$ (2.2a)

x_i are integers   (2.2b)

(This will be taken as the standard form for a *knapsack problem* for discussions in later chapters.) We treat this as a multi-stage problem by artificially insisting that the values of x_i be decided upon in the order $x_1, x_2, \ldots, x_m$. Then

stage i relates to the specification of x_i
the result of decision d_i is the value of x_i
state s_i is the time used so far = $\sum_{\alpha=1}^{i} t_\alpha x_\alpha$

(or equivalently the remaining time available), and the return $r_i(s_{i-1}, s_i)$ is the profit $p_i x_i$.

The model graph for the specific data above is shown in figure 2.12 where, for convenience, a fifth stage which yields zero return has been added. The problem clearly resolves to that of finding a longest path from vertex S to vertex F. The search tree of figure 2.13 shows that there is a unique optimal solution corresponding to the decisions

$x_1 = 0, \ x_2 = 0, \ x_3 = 1, \ x_4 = 1$

The optimal solution is $(0, 0, 1, 1)$ with value 24.

For larger problems it is impractical to draw the search tree and we must recast the method in algebraic terms for implementation on a computer. For the problem of example 2.12 define $F_i(t)$ to be the maximal total return up to and including stage i. The problem now becomes

maximise $F_n(T)$

where $F_0(0) = 0$

$$F_{i+1}(t) = \max_{0 \leq x_i \leq t/t_i} [F_i(t - x_i t_i) + x_i p_i] \quad (2.3)$$

for those values of t for which $F(t - x_i t_i)$ is defined for some $x_i \leq t/t_i$.

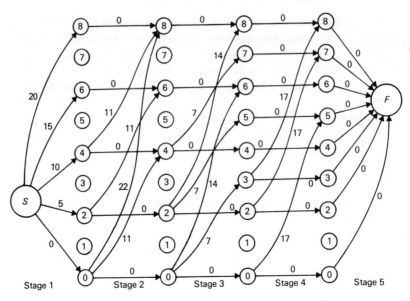

Figure 2.12 The graph associated with the knapsack problem of example 2.12

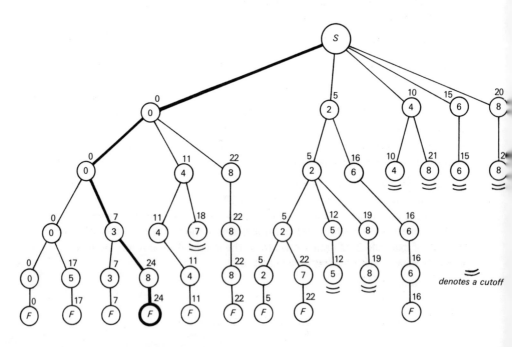

Figure 2.13 The search tree for the knapsack problem

These recursive equations are readily solved by using equation 2.3 to find $F_1(t)$ in terms of $F_0(0)$ for all relevant values of t, then $F_2(t)$ in terms of $F_1(t)$ for all relevant values of t, and so on.

Example 2.13 (Assortment problem)   An engineering firm uses rectangular cross-section bars for one of their products. This creates demand for bars with a range $w_1, w_2, \ldots, w_N$ of widths but constant depth d and length L. Demand for a non-stock width is met by cutting down a wider bar to the required size.

If n widths are to be stocked and there is a demand for v_i bars with width w_i per unit time, write down recursive equations for minimising scrap. Solve for the following data set

i	1	2	3	4	5	6	
w_i	5	7	10	14	17	22	$N = 6$
v_i	40	30	20	55	10	35	$n = 1, \ldots, 6$

Solution   We choose the ith stage to be that of deciding on the ith stock section. The state m at this stage is the index of the largest width bar that can be provided.

Let $F_i(m)$ denote the minimum scrap, using i stock sizes with widths not exceeding w_m, for satisfying all demand for sizes up to and including w_m. Then the problem becomes

minimise $F_n(N)$

with $F_0(0) = 0$, $F_0(m) = \infty$, $m > 0$

$$F_{i+1}(m) = \min_{i \leq x \leq m-1} [F_i(x) + dL \sum_{\alpha=x+1}^{m-1} (w_m - w_\alpha)v_\alpha] \quad (2.4)$$

Stage returns are given by

$$r_i(x,m) = dL \sum_{\alpha=x+1}^{m-1} (w_m - w_\alpha)v_\alpha$$

We also define $l(m) = $ a value of x leading to the minimum in equation 2.4 which allows us to construct the optimal solution.

The results of calculations for the data set above are shown in table 2.1.

Returning now to the general case suppose that it is required to minimise a separable objective function

$$\varphi = r_1(s_0, s_1) \oplus r_2(s_1, s_2) \oplus \ldots \oplus r_n(s_{n-1}, s_n)$$

for some rule of composition $\oplus$. For $i = 0, 1, \ldots, n-1$ define

$$F_{i+1}(s_{i+1}) = \min_{d_1,\ldots,d_{i+1}} [r_1(s_0,s_1) \oplus \ldots \oplus r_{i+1}(s_i,s_{i+1})]$$

$$= \min_{d_1,\ldots,d_{i+1}} g\{[r_1(s_0,s_1) \oplus \ldots \oplus r_i(s_{i-1},s_i)], r_{i+1}(s_i,s_{i+1})\}$$

Table 2.1

m \ i→	1	2	3	4	5	6
1	0 / 0	(6)	(4, 6)	(2, 4, 6)	(1, 2, 4, 6) or (2, 3, 4, 6)	(1, 2, 3, 4, 6)	(1, 2, 3, 4, 5, 6)
2	80 / 0	0 / 1				
3	290 / 0	80 / 2	0 / 2			
4	650 / 0	160 / 2	80 / 2, 3	0 / 3		
5	1085 / 0	385 / 2	160 / 4	80 / 4	0 / 4	
6	1860 / 0	700 / 4	210 / 4	130 / 4	50 / 4	0 / 5

The values of $F_i(m)$ and $l_i(m)$ are given in box (m, i) in units of dL. The optimal solutions to the problems for the case in which only i stock items are allowed are shown reading vertically

Sample calculations

$F_i(i) = 0$ (all required sizes stocked)

$$F_1(m) = F_0(0) + dL \sum_{\alpha=1}^{m-1}(w_m - w_\alpha)v_\alpha$$

$F_2(4) = \min(0 + 7 \times 30 + 4 \times 20, 80 + 4 \times 20, 290 + 0)dL = 160dL$

$l_2(4) = 2$ since minimum term is $F_1(2) + dL(w_4 - w_3)v_3$

$F_3(4) = \min(0 + 4 \times 20, 80 + 0)dL = 80dL$

$l_3(4) = 2$ or 3 as there is a tie for the minimum term

Some Basic Concepts

If g is monotonic increasing (monotonic decreasing for maximisation problems) in its first argument for each i then

$$F_{i+1}(s_{i+1}) = \min_{d_{i+1}} \{g(\min_{d_1,\ldots,d_i} [(r_1(s_0, s_1) \oplus \ldots \oplus r_i(s_{i-1}, s_i))], r_{i+1}(s_i, s_{i+1}))\}$$

$$= \min_{d_{i+1}} g[F_i(s_i), r_{i+1}(s_i, s_{i+1})]$$

$$= \min_{s_i} [F_i(s_i) \oplus r_{i+1}(s_i, s_{i+1})] \tag{2.5}$$

These recursive equations, together with initial values for $F_0(s_0)$ are sufficient to provide a solution method. Essentially, a single multi-stage problem has been replaced by a number of one-stage problems (equation 2.5). The solution itself can be found by keeping at each stage the pointers

$$l_{i+1}(s_{i+1}) = s^*$$

where s^* is a state corresponding to the minimum of equation 2.5. An example, showing the interpretation of the pointers in terms of the search tree of the problem of figure 2.10, is given in figure 4.2.

Exercise 2.7 gives examples of some objective functions which are separable and have the desired monotonicity property, thus leading to recursive equations as in equation 2.5.

EXERCISES

2.1 Show that the complete graph K_5 is non-planar by

(a) using a direct proof (that is, along lines similar to those of example 2.1);
(b) showing that Euler's formula is not satisfied.

Confirm the solution of the layout planning problem of chapter 1 by verifying that the dual graph in figure 1.3b is non-planar.

2.2 Which of the graphs of figures 1.4, 2.10, 2.12, 4.1, 5.2, 5.3, 9.1, 11.2, 11.6 are planar and which non-planar?

2.3 (Königsberg bridges problem) A *multi-graph* differs from a graph in that more than one edge is allowed between any two vertices, and an example is shown in figure 2.14. Show that the result of theorem 2.7 can be extended to multi-graphs, and hence prove that the given multi-graph cannot be traversed using each edge once and once only.

This problem is equivalent to the celebrated Königsberg bridges problem which gave rise to what is often quoted as the earliest paper on graphs (by Euler in 1736). Euler showed that it was impossible, on a walk through the town of

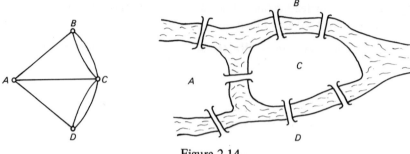

Figure 2.14

Königsberg (now Kaliningrad), to cross each of the seven bridges over the Pregel river (plan shown in figure 2.14) once and once only.

2.4 The organisers of an exhibition are considering two alternative layouts (shown in figure 2.15) both of which meet the requirements regarding display areas and shapes, demand for utilities, etc. It is intended that all visitors should follow a fixed route which passes along each line of displays just once (that is, once in each direction along avenues with displays on both sides). Moreover, it is also desirable that the stream of visitors should not 'cross itself'. Discuss the merits of the two layouts shown bearing these criteria in mind.

2.5 Devise an efficient algorithm for finding the components of a graph.

2.6 How can algorithms (K) and (P) be modified to find *maximal* spanning trees?

2.7 With the terminology of section 2.4, verify that the objective functions

(a) $r_1(s_0, s_1) \cdot r_2(s_1, s_2) \cdot \ldots \cdot r_n(s_{n-1}, s_n)$
(b) $\min [r_1(s_0, s_1), r_2(s_1, s_2), \ldots, r_n(s_{n-1}, s_n)]$
(c) $\max [r_1(s_0, s_1), r_2(s_1, s_2), \ldots, r_n(s_{n-1}, s_n)]$

are separable and satisfy the monotonicity property. Write down corresponding recursive equations.

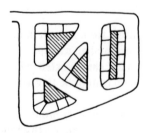
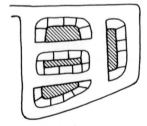

Figure 2.15

Some Basic Concepts

2.8 Using the given distance matrix find

$$\begin{bmatrix}
- & 22 & 61 & 60 & 60 & 57 & 69 \\
22 & - & 37 & 58 & 40 & 38 & 51 \\
61 & 37 & - & 62 & 15 & 39 & 19 \\
60 & 58 & 62 & - & 43 & 23 & 53 \\
60 & 40 & 15 & 43 & - & 25 & 17 \\
57 & 38 & 39 & 23 & 25 & - & 31 \\
69 & 51 & 19 & 53 & 17 & 31 & -
\end{bmatrix}$$

(a) a minimal spanning 3-forest;
(b) a minimal spanning 2-forest;
(c) an MST

Do the items fall naturally into two or three clusters?

2.9 By developing a search tree, as in figures 2.11 and 2.13, solve the production planning problem of chapter 1 (p. 7) and check your answer against figure 1.4.

By choosing a suitable definition for 'state' and the function F_i write down recursive equations for solving this problem.

2.10 A control system consists of n subsystems $S_1, S_2, \ldots, S_n$. All S_i must be functioning for the whole system to function. S_i can be duplicated with the cost of adding the standby unit being c_i. The probability that S_i fails is p_i (as is the probability that the standby to S_i fails). If a total budget C is available for standby units, which units should be duplicated so as to maximise the system reliability?

Let $F_i(x)$ denote the probability that S_j, or its standby, is functioning, $j = 1, \ldots, i$, if a total x of the budget has been allocated. Then

$$F_i(x) = \max_{\substack{t=0,1 \\ tc_i \leq x}} [F_{i-1}(x - tc_i)(1 - p_i^{t+1})], \quad F_0(0) = 1$$

Clearly the problem becomes that of finding $F_n(C)$. Solve for the data following

i	1	2	3	
c_i	3	4	5	$C = 7, 8, 9$
p_i	0.1	0.2	0.3	

2.11 [*Media scheduling – Zufryden (1975)*] An advertiser must select times for advertisement insertions from a set $\{T_\alpha\}$, $\alpha = 1, 2, \ldots, n$, of possible times.

He wishes to do this so as to maximise sales, subject to an overall budget limit K, over a planning period (τ_0, τ_n) where $\tau_0 < \tau_1 < \ldots < \tau_{n-1} < \tau_n$.

$\mu(t) + \mu_\infty$ is the rate of sales at time t, where μ_∞ is the rate that could be maintained in the absence of advertising, and $F_i = \int_{\tau_0}^{T_i} \mu(t)\,dt$ is the total extra sales to time τ_i. The function μ is discontinuous at $t = \tau_i$ if an insertion occurs

$$\mu(\tau_i + 0) = [A + B\mu(\tau_i - 0)]\,v(\tau_i) + \mu(\tau_i - 0)\,[1 - v(\tau_i)]$$

where A and B are constants, $v(\tau_i)$ is the probability of a person seeing an advertisement exposed at time τ_i, and

$$\mu(\tau_i - 0) \equiv \lim_{\epsilon \to 0} \mu(\tau_i - \epsilon),\quad \mu(\tau_i + 0) \equiv \lim_{\epsilon \to 0} \mu(\tau_i + \epsilon)$$

μ is assumed to decay exponentially between insertions

$$\mu(\tau_i + t) = \mu(\tau_i + 0)e^{-\beta t} \qquad \beta > 0$$

Let

$$f_i = \int_{T_i+0}^{T_{i+1}-0} \mu(t)\,dt = \mu(\tau_i + 0)\lambda_i,\quad \lambda_i = (1 - e^{-\beta(\tau_{i+1} - \tau_i)})/\beta$$

denote the sales in period (τ_i, τ_{i+1}). Then it can easily be seen that f_i depends not only on the accumulated cost of insertions but also on the particular insertions made. That is the effectiveness of an insertion at time τ_i depends explicitly on prior insertion decisions.

What would be an appropriate definition for a state of the system if a dynamic programming solution were contemplated? Write down recursive equations for solving this problem.

2.12 Suppose, for exercise 2.11, that $B = 1$; then if an insertion occurs at τ_i,

$$\mu(\tau_i + 0) = \mu(\tau_i - 0) + Av(\tau_i)$$

Show that if $W_j = Av(\tau_j)(1 - e^{-\beta(T_n - T_j)})/\beta$ then

$$F_n - \mu(\tau_0 + 0)(1 - e^{-\beta(T_n - T_0)}) = \sum_{j=1}^{n-1} W_j \xi_j$$

where $\xi_j = 1$ if an insertion occurs at τ_j and 0 otherwise. Show that the problem of maximising F_n can thus be modelled as a knapsack problem.

3 Branch-and-bound Methods

As pointed out in section 2.4, multi-decision problems are represented naturally by trees in which a node x represents the set of (component) decisions actually made, and the branches from x represent the possible 'choices of action at x'. The problem of finding an optimal composite decision is equivalent to the problem of finding an optimal path from the root to a terminal node in a search tree (see section 2.4), and one technique for achieving this aim is that of discrete dynamic programming as shown in chapter 2. In this chapter an alternative technique, called branch-and-bound, is introduced and described.

Branch-and-bound (abbreviation B & B) concepts are introduced informally by means of an application to a ranking problem, then the underlying philosophy is discussed. One of the earliest applications of B & B was in the field of integer linear programming and section 3.2 gives an introduction by means of a simple example which is solved using Dakin's method. However, details of the theory of integer linear programming are beyond the scope of the present work and for an assessment of the importance of B & B techniques in this area interested readers may consult Garfinkel and Nemhauser (1972), Salkin (1975) or Taha (1975).

Separate techniques have been developed for the special case of problems in 0–1 variables. A discussion of such techniques and illustrative problems is given in section 3.2.

3.1 CONCEPTS OF B & B

B & B techniques are frequently useful when a function, f say, is to be minimised or maximised over a finite set F of alternatives. (For definiteness the present discussion will be restricted to finding *an* optimal solution to minimisation problems.) Complete enumeration trees can become very large even for quite small multi-decision problems. For example, a problem requiring 8 component decisions each with a choice from 7 actions, involves a tree with 7^8 terminal nodes, that is, well over 5 000 000 possible composite decisions. Usually a multi-decision problem is less orderly but often it is a straightforward matter to regard such a problem as a d-decision–c-choice one. Generally, if the specification of the set F of *feasible solutions* is such that the generation of elements of F is 'awkward' it is often convenient to start with a larger, though still finite, set C of *candidate solutions* discarding, as they arise, sets containing only infeasible solutions. An example of this is given in the second treatment of the ranking problem below.

Example 3.1 The preferences expressed in a paired comparison experiment over the set $\{a,b,c,d\}$ are given by the following matrix, and it is required to find a ranking which minimises the number of discrepancies (that is, the number of 'violated' preferences in which x is ranked above y but y is preferred to x).

	a	b	c	d	Total score
a	·	1	1	0	2
b	0	·	0	1	1
c	0	1	·	1	2
d	1	0	0	·	1

Solution (1) This problem will be treated as in Burstall (1967), the decisions being explained as they are encountered. Now a has a better score than b and also a is preferred to b, so the first decision that will be presented is, should a be ranked above b or not? This is a reasonable question to pose, since it is seems likely that any optimal ranking will have a ranked above b and so it may be hoped that any rankings with b ranked above a need not be considered further.

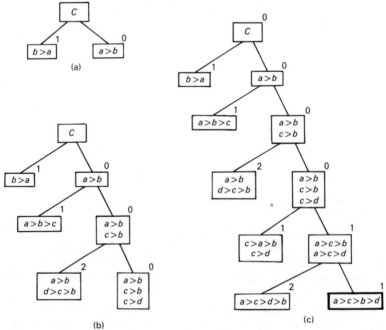

Figure 3.1 Three stages in the growth of the search tree for solution (1) of the ranking problem. Lower bounds are shown beside each node

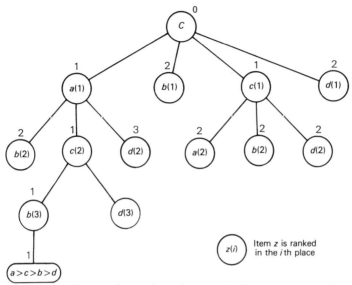

Figure 3.2 The search tree for solution (2) of the ranking problem

This hope is entirely lived up to, though in other cases we might not be so fortunate. None the less, such a strategy should tend to be favourable generally.

Of the two nodes added to the search tree the one with $a > b$ (that is a ranked above b) is the more promising since $a > b$ does not (immediately) imply any discrepancies. From the preference matrix it is seen that c is also noticeably better than b and the second branching corresponds to whether or not $c > b$.

After three further branchings the search tree developed is as shown in figure 3.1c. Since the ranking $a > c > b > d$ involves only one discrepancy, and any solution corresponding to the solution subset at any other leaf node implies at least one discrepancy, $a > c > b > d$ must be an optimal ranking. Of course it has not been proved at this stage that this is the only optimal solution. To establish this or to find the other optimal solutions, it is necessary to develop all leaf nodes with only one implied discrepancy.

The above method led easily to a solution for this example, this being partly due to judicious choices for branching (see exercise 3.1).

Solution (2) Let $x_1 > x_2 > x_3 > x_4$ be a ranking, then x_i, $i = 1, \ldots, 4$ can be any element of the set $\{a,b,c,d\}$. There are $4^4 = 256$ different quadruples (x_1, x_2, x_3, x_4) of which only $4! = 24$ correspond to valid rankings. That is, of the 256 candidate solutions 232 are infeasible and these must be eliminated. The branching rule that will be adopted is as follows: the first branching corresponds to the choice of x_1, the top ranked element, the second and third branchings correspond to the choices of x_2 and x_3 respectively. (Of course if x_1, x_2 and x_3 are specified then x_4 is automatically determined.) The search tree is developed in a similar way to that of the first solution method and is shown in figure 3.2.

As before, $a>c>b>d$ is found to be optimal.

While a node $\boxed{a(2)}$ is a potential successor to the node $\boxed{a(1)}$, this is not included in the tree since it contains infeasible solutions only. Other such sets are also rejected.

In this case both solution methods led directly to an optimal ranking. However, in general this is not the case and a certain amount of 'backtracking' will be needed.

It will be assumed in the remainder of this chapter that all partitions are proper. Branching starts by partitioning C into disjoint sets $X_1, \ldots, X_r$ and then if there is some relatively easy way of telling that *an* optimal solution lies in X_k, for some k, the problem has been reduced to that of minimising f over X_k. The branching continues by *developing* X_k, that is partitioning it into subsets $X_{k_1}, \ldots, X_{k_s}$. Again if it can be ascertained that an optimal solution lies in X_{k_l} say, then the problem has been further reduced, and so on. This is the theme behind B & B techniques, namely that whole subsets of solutions can be discarded and the problem reduced to smaller and smaller problems whose optimal solutions provide an optimal solution to the original problem. In general it is not sufficient to consider a single subset resulting from each branching, and backtracking becomes necessary.

The branching and consequent reductions of the problem can be represented by means of a search tree (section 2.4). Each node of the tree corresponds to a subset of C itself. C is the first node to be developed and nodes $X_1, \ldots, X_r$ and edges $CX_1, \ldots, CX_r$ are added. Other nodes are developed and further nodes and edges added. In this way B & B search trees are *grown* (figure 3.1 for example).

Branching Rules

If branching were to be continued for long enough then, since C is finite and all partitions proper, a stage would be reached at which a complete enumeration tree would have been formed.

Example 3.2 A permutation (ranking, sequencing, ordering) problem can be solved by assigning a rank r_i to each of the n items of the set $A = \{a_1, \ldots, a_n\}$. The set of feasible solutions can be taken to be the set of n-dimensional vectors

$$F = \{x \mid x_i \in A \ \& \ x_i \neq x_j \text{ if } i \neq j\}$$

where x_i is the item with rank i. Discuss possible branching rules and form corresponding complete enumeration trees for the case of $n = 4$ and $A = \{a, b, c, d\}$.

Solution Three natural branching rules (that have been used in practice) will be described briefly.

(1) Rank branching Let $R = \{r_1, \ldots, r_p\}$ be a subset of $\{1, 2, \ldots, n\}$ containing p distinct elements. A node X of the search tree is determined by imposing the constraints

$$x_{r_1} = a_{j_1}, \quad x_{r_2} = a_{j_2}, \ldots, \quad x_{r_p} = a_{j_p}$$

for some set $\{a_{j_1}, \ldots, a_{j_p}\}$ of p distinct items. X can be partitioned into $n-p$ disjoint subsets X_i by specifying the item to have rank $r(X)$ for some $r(X) \notin R$; that is

$$X_k = \{x \mid x \in X \ \& \ x_{r(X)} = a_k, k \neq j_1, \ldots, j_p\}$$

This branching rule will be termed *rank branching*, and in the general form given requires the use of an explicit *selection rule*, for determining $r(X)$ when X is given. However, in practice $r(X)$ is taken to be a function only of the *level* of X (that is, p). For example, $r(X) = p + 1$ was used in solution (2) of example 3.1 and $r(X) = n - p$ is sometimes used.

Figure 3.3a shows a complete enumeration tree, for $n=4$ and $A = \{a, b, c, d\}$ generated by rank branching. The subset of solutions corresponding to a particular node will depend on the selection rule used! For example, consider the nodes marked U, V, W and Y at levels 0, 1, 2 and 3 respectively. Then Y corresponds to the single solution

(b, d, a, c)	if	$r(U) = 3, r(V) = 1$ and $r(W) = 4$
(b, a, d, c)	if	$r(U) = 2, r(V) = 1$ and $r(W) = 4$
(a, b, c, d)	if	$r(X) = $ (level of X) $+ 1$ for all X
(d, c, b, a)	if	$r(X) = n - $ (level of X) for all X

(2) Immediate successor branching A node X is specified by p (= level of X) relations of the form $r_j = r_i + 1$ (to be interpreted as item a_j is ranked *immediately* after item a_i). X may now be partitioned into two sets

$$X(kl) = \{x \mid x \in X \ \& \ r_l = r_k + 1\}$$
$$X(\overline{kl}) = X - X(kl) = \{x \mid x \in X \ \& \ r_l \neq r_k + 1\}$$

A selection rule is required to decide on the pair of indices k and l.

Figure 3.3b gives an example of a complete enumeration tree for $n = 4$ and $A = \{a, b, c, d\}$.

(3) Successor branching A node X is specified by a set of p relations $r_j > r_i$ (to be interpreted as item a_j is ranked after a_i). X may be partitioned into two sets

$$X(kl) = \{x \mid x \in X \ \& \ x_l > x_k\}$$
$$X(lk) = X - X(kl) = \{x \mid x \in X \ \& \ x_k > x_l\}$$

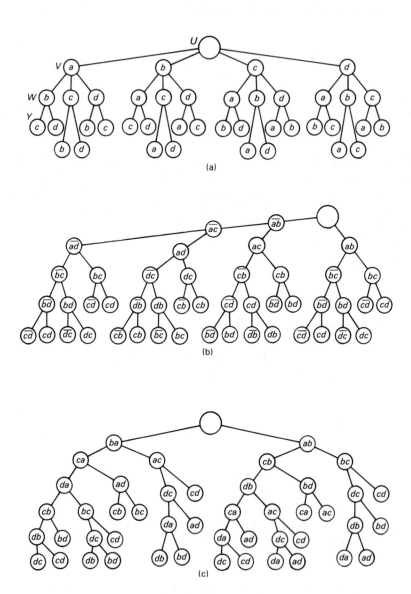

Figure 3.3 Complete enumeration trees for a permutation problem involving four objects and with (a) rank; (b) immediate successor and (c) successor branching strategies

Again a selection rule is required to decide on the pair (k, l) $k \neq l$. Figure 3.3c gives an example of a complete enumeration tree. Successor branching was used in solution (1) of example 3.1.

Bounding Rules

A search tree should form but a very small part of a complete enumeration tree because of the enormous size of the latter. A means of achieving this is provided by a suitably chosen *bounding rule*. A *bounding function* is a function $g : 2^C \to R$, which assigns a real number $g(x)$ to each subset of C such that if $X \subset C$ then the bound $g(X)$ satisfies

$$g(X) \leq f(x) \quad \text{all } x \in X \cap F$$

(Note that the bounds are all *lower* bounds since we are dealing with minimisation problems.) It will also always be assumed that

$$g(\{x\}) = f(x) \quad \text{all } x \in F \tag{3.1}$$

Bounds for a problem P can be obtained by relaxing some of the constraints. This leads to a *relaxed* problem Q whose feasible set F_Q contains the feasible set F_P of P. Let X be a subset of C then

$$g(X) = \min_{y \in X \cap F_Q} f(y) \leq \min_{y \in X \cap F_P} f(y) \leq f(x) \quad \text{all } x \in X \cap F_P$$

and g is clearly a lower bounding function.

Example 3.3 (Linear assignment problem; see sections 7.1 and 10.3) Four men are available to perform four tasks. Suppose man i has a rating of u_{ij} as to the undesirability of his performing task j and that it is required to find an assignment which minimises the total of the ratings. Obtain a simple bounding function and illustrate using the data set

	1	2	3	4	row minimum
1	15	5	9	5	5
2	17	14	4	7	4
3	9	13	13	7	7
4	13	10	6	15	6

Solution Let

$$C = \{x \mid x = (i, j, k, l) \ \& \ i, j, k, l \in \{1, 2, 3, 4\}\}$$

be the set of candidate solutions. The (linear) assignment problem P is to minimise $\varphi = u_{1i} + u_{2j} + u_{3k} + u_{4l}$ where (i, j, k, l) is a permutation of $(1, 2, 3, 4)$.

A relaxed problem Q is that for which the constraint is relaxed to $i,j,k,l \in \{1,2,3,4\}$. Clearly $F_Q = C$ in this case. Examples of bounds are

$$g(C) = \min_{1 \leq \alpha \leq 4} u_{1\alpha} + \min_{1 \leq \beta \leq 4} u_{2\beta} + \min_{1 \leq \gamma \leq 4} u_{3\gamma} + \min_{1 \leq \delta \leq 4} u_{4\delta} = 22$$

$$g(C[i=3]) = u_{13} + \min_{1 \leq \beta \leq 4} u_{2\beta} + \min_{1 \leq \gamma \leq 4} u_{3\gamma} + \min_{1 \leq \delta \leq 4} u_{4\delta} = 26$$

$$g(C[i=3],[j=2]) = u_{13} + u_{22} + \min_{1 \leq \gamma \leq 4} u_{3\gamma} + \min_{1 \leq \delta \leq 4} u_{4\delta} = 36$$

where $C[i=3]$ denotes the subset of C with i fixed at 3, etc.

It is readily seen that if i is set at 3 then we may insist that $\beta, \gamma, \delta \neq 3$. This leads to the improved bounding function g'

$$g'(C) = g(C) = 22$$

$$g'(C[i=3]) = u_{13} + \min_{\beta=1,2,4} u_{2\beta} + \min_{\gamma=1,2,4} u_{3\gamma} + \min_{\delta=1,2,4} u_{4\delta} = 33$$

$$g'(C[i=3],[j=2]) = u_{13} + u_{22} + \min_{\gamma=1,4} u_{3\gamma} + \min_{\delta=1,4} u_{4\delta} = 43$$

Note that g' satisfies equation 3.1 without modification. Note also that $g'(X) \geq g(X)$ for all $X \subset C$; g' will be said to be *stronger* or *tighter* than g.

Let ξ be the *incumbent* solution, that is a best solution so far found. If $f(\xi) \leq g(Y)$ for some node Y (that is $Y \subset C$) then

$$f(\xi) \leq g(Y) \leq f(x) \quad \text{all } x \in Y \cap F \tag{3.2}$$

and so Y can be eliminated from further consideration since it cannot possibly contain a solution *better* than ξ. That is, node Y becomes *inactive* and need not be developed thus allowing a reduction in the size of the B & B tree grown. However, it could have happened that node Y was developed before solution ξ was discovered and the saving mentioned above not (wholly) achieved. Thus it is clear that, in general, the extent to which the B & B tree is grown depends on the order in which nodes are developed, and the *development rule* or *strategy* can critically affect the usefulness of the technique.

Development Rules

(1) Depth-first A common development rule is the *depth-first* (DF) strategy of branching from a most recently created active node (a node Y is *active* as long as $f(\xi) > g(Y)$ where ξ is the incumbent solution). This rule forces the tree to a terminal node quickly. Although there might be more computation required through there being fewer cutoffs, it requires a minimal amount of storage, since it is sufficient to store problems corresponding to each node on one route from the root of the B & B tree to a terminal node.

(2) Breadth-first (BF) This is the development rule of dynamic programming and is the opposite of depth-first in that the strategy is to branch from a 'least recently created' active node.

(3) Uniform-cost This rule consists of choosing a leaf node X, of the partially grown tree, for which

$$g(X) \leqslant g(Y) \quad \text{all leaf nodes } Y$$

ties being resolved by some subsidiary rule. This is the 'branch from lowest bound' or *uniform-cost* (UC) strategy and is computationally optimal in the sense that it effectively requires a minimal number of nodes to be developed. However, although leading to small trees, UC might fail to lead to minimal computing time and also suffers the defect of first producing a feasible solution at or near the end of the calculation.

The three strategies described above are shown in figure 3.4.

There are various other strategies that may be used (Benichou *et al.*, 1971) including compromises between uniform-cost and depth-first. In these, development is depth-first until backtracking must take place when a node with least lower bound is developed next: this is a *change when necessary* strategy.

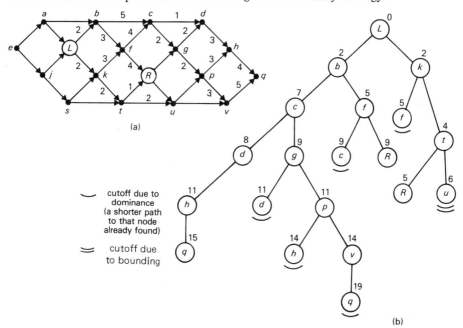

Figure 3.4 Search trees for finding a shortest path from L to R in the graph of (a) are shown in (b), (c) and (d) using depth-first, breadth-first and uniform-cost development strategies respectively.

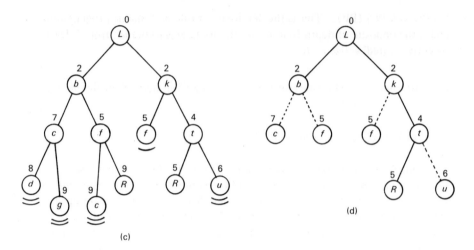

Figure 3.4 *continued*

It may be seen that the more closely g approximates the *greatest* lower bounding function g^* the more cutoffs are likely to be achieved. Despite the fact that a good approximation is often difficult and time consuming to evaluate, its use, nevertheless, usually leads to less overall computation. A desirable property of g is that $g(A) \geq g(B)$ if $A \subset B$; if this is not so then a stronger bounding function g'

$$g'(A) = \max\,[g(A), g'(P)]$$

may be used where A arises by developing node P. Again, if a variety of rules $g_1, \ldots, g_r$ are available then the (in general) stronger rule $g = \max_i g_i$ may be used (see exercises 3.5 and 3.6).

Dominance

In some applications of B & B it may happen that there are ways, outside the framework of B & B, which allow one to state that for some pairs of sets $X, Y \subset C$ the best solution in X is at least as good as the best solution in Y. X is said to *dominate* Y and the node Y of the B & B tree need not be developed. (If X dominates Y and Y dominates X then either may be made inactive.) For an example of the use of dominance see exercise 3.6.

Finally, it may be noted that B & B techniques are exact in that they guarantee optimal solutions. However, they usually still involve a considerable amount of computation and if a good suboptimal solution is acceptable then it is often better to use B & B 'heuristically' in that the best feasible solution generated, in a given time say, is retained (chapter 11).

3.2 INTEGER LINEAR PROGRAMMING

One of the earliest applications of a B & B technique was that of Land and Doig (1960) to mixed integer linear programming (MILP). The problem can be formulated as

$$\text{minimise} \quad \varphi = \sum_{j=1}^{n} c_j x_j \qquad (3.3a)$$

subject to

$$\sum_{j=1}^{n} a_{ij} x_j \geq b_i \quad i = 1, 2, \ldots, m \qquad (3.3b)$$

$$x_j \geq 0 \quad j = 1, 2, \ldots, n \qquad (3.3c)$$

$$x_1, x_2, \ldots, x_s \text{ are integers for } s \leq n \qquad (3.4)$$

If $s = 0$ (that is, constraint 3.4 does not apply) then this is just an ordinary linear programming (LP) problem for which effective algorithms, of the simplex type, exist. However, the presence of integrality constraints considerably increases the complexity, and B & B methods are at present among the best available. If $s = n$ in constraint 3.4 then the problem is a pure integer linear programming (ILP) one and is essentially combinatorial in nature. Though we only deal with ILPs, Dakin's method used below is equally applicable to the mixed integer case.

Of course equation 3.3 could be solved as an LP and if an optimal solution x^* were to be obtained with $x_1^*, \ldots, x_s^*$ integral then x^* would also be an optimal solution of the corresponding MILP or ILP. At any rate it seems reasonable to investigate solutions 'near' to x^* with at least some integrality conditions imposed. Dakin's (1965) method is essentially a systematic way of achieving this.

Example 3.4

$$\text{Minimise} \quad \varphi = 4x + 9y \qquad (3.5a)$$

subject to

$$2x + 5y \geq 10 \qquad (3.5b)$$

$$2x + y \geq 4 \qquad (3.5c)$$

$$x, y \geq 0 \qquad (3.5d)$$

$$x \text{ and } y \text{ are integral} \qquad (3.6)$$

Solution The space of feasible solutions of problem 3.5 and 6 is shown in figure 3.5.

The solution to the LP problem 3.5 is $(5/4, 3/2)$ with $\varphi = 18\frac{1}{2}$.

This is not an integer feasible solution but does provide a lower bound to the

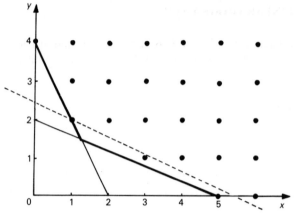

Figure 3.5

optimal integer solution. Now $x = 5/4$ is not feasible but we can assert that $x \leqslant 1$ or $x \geqslant 2$ and so two new problems are created by adding respectively the constraints $x \leqslant 1$ and $x \geqslant 2$ to those of 3.5; that is, we branch from the root node C as illustrated in figure 3.6. (In general a variable x_i, which is required to be integral but assumes a non-integer value $\tilde{x}_i$ in some LP solution, is selected. Then new problems are formed by adding constraints $x_i \leqslant \lfloor \tilde{x}_i \rfloor$ and $x_i \geqslant \lceil \tilde{x}_i \rceil$, where '$\lfloor \ \rfloor$' and '$\lceil \ \rceil$' respectively denote 'the largest integer not greater than' and 'the smallest integer not less than'. A good selection rule for specifying the branching variable x_i is important as regards computational efficiency of the method; a little is said about this below.)

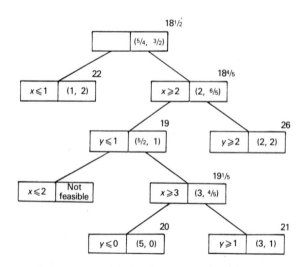

Figure 3.6 B & B search tree for example 3.4

The solutions to the two new problems are (1,2) with $\varphi = 22$ and (2, $6/5$) with $\varphi = 18^4/5$. The latter node looks more promising and is selected for development by introducing new constraints $y \leqslant 1$ and $y \geqslant 2$. Solutions ($5/2$, 1) with $\varphi = 19$ and (2,2) with $\varphi = 26$ are obtained. The latter node can be made inactive since we have already found a better feasible solution [$\varphi = 22$ at (1,2)]. The former node is now developed by introducing new constraints $x \leqslant 2$ and $x \geqslant 3$. However the constraint $x \leqslant 2$ together with the earlier $x \geqslant 2$ (introduction of constraints is cumulative) implies $x = 2$ and, since $y \leqslant 1$, leads to no feasible solutions. The solution continues as shown in figure 3.6, and $(x,y) = (5,0)$ is found to be optimal with value 20.

The B & B tree may seem large for the size of problem involved but this is at least partly due to the fact that the problem was constructed for illustrative purposes with the optimal solution relatively far away from the optimal LP solution (figure 3.5).

The selection rule for choosing the branching variable can considerably affect the computational efficiency of the method. (For example, it is easily verified that branching on y before x leads to a smaller B & B tree for the above example.) Benichou *et al.* (1971) select that variable which leads to the largest 'expected' increase in lower bounds (for minimisation problems) thus increasing the chances of obtaining cutoffs and so tending to lead to smaller search trees. Consider a node P of the search tree for which $\tilde{x}_k$ is non-integer, and the successor nodes L and R obtained by imposing the constraints $x_k \leqslant \lfloor \tilde{x}_k \rfloor$ and $x_k \geqslant \lceil \tilde{x}_k \rceil$ respectively. Let the lower bounds at P, L and R be $g(P), g(L)$ and $g(R)$. The *down* and *up pseudocosts* are

$$\text{pcd}(k) = \frac{g(L) - g(P)}{\tilde{x}_k - \lfloor \tilde{x}_k \rfloor}$$

$$\text{pcu}(k) = \frac{g(R) - g(P)}{\lceil \tilde{x}_k \rceil - \tilde{x}_k}$$

A branching variable can now be chosen as one for which the *pseudocost*

$$PC(k) = \min \{ \text{pcd}(k)(\tilde{x}_k - \lfloor \tilde{x}_k \rfloor), \text{pcu}(k)(\lceil \tilde{x}_k \rceil - \tilde{x}_k) \}$$

is maximal. As it stands the idea of pseudocost is of no help. However, the selection rule is based on the use of $PC(k)$ with estimates for $\text{pcd}(k)$ and $\text{pcu}(k)$, based on statistics gained from earlier branching [see Benichou *et al.* (1971) for further details]. Pseudocosts were found to be very promising for ILPs generally by Benichou *et al.*, but Breu and Burdet (1974) found them to be less effective for the special class of 0–1 problems to which we now turn.

Zero–One Programming

A special type of ILP problem is one in which all variables are restricted to the values 0 and 1. Such problems can be put into the form

$$\text{minimise} \quad \varphi = \sum_{j=1}^{n} c_j x_j \quad c_j > 0 \quad (3.7a)$$

$$\text{subject to} \quad \sum_{j=1}^{n} a_{ij} x_j \leq b_i \quad i = 1, 2, \ldots, m \quad (3.7b)$$

$$x_j = 0 \text{ or } 1, \quad j = 1, 2, \ldots, n \quad (3.7c)$$

Let E and I be subsets of $X = \{1, 2, \ldots, n\}$ corresponding to the sets of variables assigned the values 0 and 1 respectively and let $U = X - (E \cup I)$. For the set of feasible solutions corresponding to a particular node, E and I give the variables which are respectively *E*xcluded from and *I*ncluded in the set of variables which assume the value 1; U gives the set of *U*nassigned variables. Each node of the B & B tree corresponds to a partition (E, U, I) with the root node corresponding to $(\phi, \{1, \ldots, n\}, \phi)$. One way of branching consists of selecting a variable x_k with $k \in U$ and assigning it to E or I as in figure 3.7. Assuming the ILP problem has been formulated so that $c_i > 0$ for all i (this can always be done) then, if P is the node (E, U, I)

$$l(P) = \sum_{j \in I} c_j$$

is a lower bound to any feasible solution in P. This simple scheme provides a basis for a B & B solution to 0–1 problems.

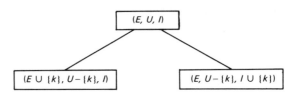

Figure 3.7

In order to cut down the size of the search tree a test for feasibility can be made at each node (Balas, 1965) by checking each constraint individually. Several such tests have been devised (Taha, 1975).

Consider the ith constraint and let

$$U^- = \{j \mid j \in U, a_{ij} < 0\}, \quad U^+ = U - U^- \text{ and } d_i = b_i - \sum_{j \in I} a_{ij}$$

Then substituting the values of the variables fixed at the node under consideration, gives

$$\sum_{j \in U} a_{ij} x_j \leq b_i - \sum_{j \in I} a_{ij} = d_i \quad (3.8)$$

If $d_i \geq 0$ this inequality can be satisfied trivially. If $d_i < 0$ then a check can be made as to whether $x_j, j \in U$ can be chosen so that the inequality is satisfied, that is, whether

$$\sum_{j \in U^-} a_{ij} \leq d_i$$

If this is not true then the node (E, U, I) cannot lead to any feasible solutions and a cutoff is obtained.

Test 0 (Fix at zero) Sometimes the inequality 3.8 may be satisfied but only if some variable assumes a particular value. If $x_k = 1$, $k \in U^+$ we require (from inequality 3.8) that

$$\sum_{j \in U^-} a_{ij} \leq d_i - a_{ik}$$

and hence if this is not satisfied then x_k must be 0 at node (E, U, I) for feasibility to be achieved. The test would naturally be made for that $k \in U^+$ for which a_{ik} is maximal.

Test 1 (Fix at one) If $x_k = 0$, $k \in U^-$ we require (from inequality 3.8) that

$$\sum_{j \in U^-} a_{ij} \leq d_i + a_{ik}$$

and hence if this is not satisfied x_k must be 1 at node (E, U, I) for feasibility to be achieved. The test would naturally be made for that $k \in U^-$ for which a_{ik} is minimal.

Despite improved feasibility checks the above scheme is not very efficient and improved bounds are required. The LP relaxation, with $x_j \in \{0, 1\}$ replaced by $0 \leq x_j \leq 1$ all j, can be used as described earlier in this section. This opens the way for the use of penalties and pseudocosts.

Knapsack Problem

Example 3.5 (Kolesar, 1966)

maximise $\quad 11x_1 + 9x_2 + 8x_3 + 15x_4$

subject to $\quad 4x_1 + 3x_2 + 2x_3 + 5x_4 \leq 8$

$\quad x_1, x_2, x_3, x_4 \in \{0, 1\}$

Solution Upper bounds (of relevance since this is a maximisation problem) will be obtained by relaxing the constraint $x_j \in \{0, 1\}$ to $0 \leq x_j \leq 1$. Now the solution of the continuous knapsack problem

$$\text{maximise} \quad \varphi = \sum_{i=1}^{n} p_i \xi_i \tag{3.9a}$$

subject to $\sum_{i=1}^{n} t_i \xi_i \leq T$ (3.9b)

$0 \leq \xi_i \leq 1, \quad i = 1, \ldots, n$ (3.9c)

in which T and all t_i are strictly positive, is easily obtained. Assuming, without loss of generality, that the variables ξ_i are so indexed that

$$\frac{p_1}{t_1} \geq \frac{p_2}{t_2} \geq \ldots \geq \frac{p_n}{t_n}$$ (3.10)

then an optimal solution is

$\xi_1 = \xi_2 = \ldots = \xi_l = 1$ (3.11a)

$\xi_{l+1} = (T - \sum_{i=1}^{l} t_i)/t_{l+1}$ (3.11b)

$\xi_{l+2} = \ldots = \xi_n = 0$ (3.11c)

where l is the first index for which the right hand side of equation 3.11b does not exceed 1.

Upper bounds can now be obtained at any node (E, U, I) by using inequality 3.10 and equations 3.11 applied to the variables $x_i, i \in U$. Now $8/2 > 9/3 = 15/5 > 11/4$ and the lower bound at the top node is $8 + 9 + (3/5)15 = 26$ corresponding to the solution $x_3 = 1, x_2 = 1, x_4 = 3/5, x_1 = 0$. Now branch on x_3. Setting $x_3 = 0$, the bound is $9 + 15 = 24$ corresponding to solution $x_2 = 1, x_4 = 1, x_1 = 0$. This together with $x_3 = 0$ gives a feasible solution to the original problem. It will then be a best feasible solution corresponding to this node (and in fact the incumbent solution). Consequently node 2 (figure 3.8) need not be developed. Backtracking to node 3, the solution is clearly the same as for the root node, and so nodes 4 and 5 are created. The solution to the bounding problem at node 4 is $x_4 = 1, x_1 = 1/4$ giving a bound of $8 + 15 + (1/4)11 = 25\frac{3}{4}$ (or 25 if it is noted that the optimal solution must be integer valued).

The solution continues as shown in figure 3.8. The cutoff at node 5 is obtained by noting that setting $x_3 = 1$ and $x_2 = 1$ uses up $t_3 + t_2 = 5$ units of the 'knapsack' T leaving 3 units remaining. But $t_1, t_4 > 3$ and so $x_1 = x_4 = 0$ in all feasible solutions corresponding to node 5. Consequently the bound is set at 17 ($=p_3 + p_2$) and a cutoff is obtained. Similar remarks apply to node 7.

Branching Strategies for 0–1 Problems

There are three basic branching strategies that may be used for 0–1 problems which will be termed *inclusion/exclusion, inclusion* and *exclusion strategies*. In the first strategy branching is binary and depends, at any node, on whether the branching variable is to be set to 0 or 1. With an inclusion (exclusion) strategy branching, at any node, is on the variable in the corresponding unassigned set U which is to be set to 1 (0). The differences are most easily seen from an

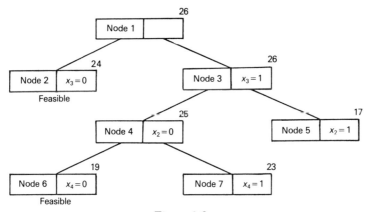

Figure 3.8

example in which, for every feasible solution, the number of variables taking the value 1 is fixed.

Example 3.6 (Facility location) Two facilities are to be sited at locations chosen from six possible locations so as to minimise some objective function φ. (For the form that φ might take the reader is referred to chapter 5.) Determine complete enumeration trees for the three basic branching strategies.

Solution The locations are numbered 1 to 6 and variables x_i, $i = 1,\ldots,6$ introduced where $x_i = 1$ if a facility is to be sited at location i and $x_i = 0$ otherwise. The three trees are illustrated in figure 3.9. In order that the trees be non-redundant it is necessary that partitioning should be proper. For inclusion/exclusion branching this condition requires that a node be regarded as terminal if either two locations have been decided upon or four locations have been ruled out.

For inclusion branching care must be taken to avoid more than one node corresponding to a particular subset of solutions. Thus, in this example, siting a facility at location j then one at location k must not be regarded as being different from siting a facility at location k then one at location j. This can be achieved for a particular indexing of the variables by regarding the set with $x_k = 1$ as implying $x_j = 0$ for all previously unassigned variables x_j, $j < k$. Notice that in figure 3.9b the node 5, created at the first branching, need not be developed as $x_5 = 1$ implies $x_1 = x_2 = x_3 = x_4 = 0$ and, since two facilities are to be sited, x_6 must be 1. Notice also that a node 6 resulting from the first branching could not contain any feasible solution and is not shown in the figure.

Exclusion branching is formally equivalent to inclusion branching, one being converted into the other by changing to variables y_i, where $y_i = 1 - x_i$ all i. Thus for the problem under consideration determining the locations for two facilities is the same as determining the four locations that will not be assigned a facility.

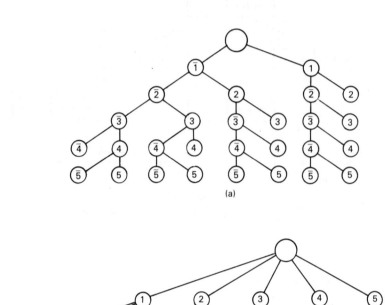
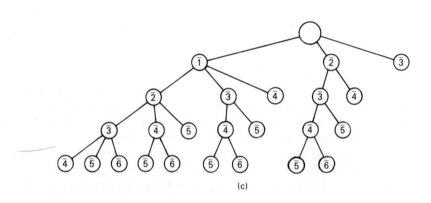

Figure 3.9 Search trees illustrating (a) inclusion/exclusion; (b) inclusion and (c) exclusion branching strategies

As seen from figure 3.9, the inclusion/exclusion complete enumeration tree has more nodes than the others. However, it does not follow that this is true for the B & B trees that result from inclusion/exclusion branching. Indeed the trees may well be smaller than the corresponding trees for inclusion and exclusion branching if a good selection rule for choosing the branching variable is available.

Other schemes for 0–1 ILPs are given in Lawler and Bell (1967) and Beale (1970).

EXERCISES

3.1 Solve the ranking problem of example 3.1 using solution method (1) but with the following selection rule: branching from levels 0,1,2,3 and 4 is on whether $(a > d$ or $d > a)$, $(b > c$ or $c > b)$, $(a > c$ or $c > a)$, $(a > b$ or $b > a)$ and $(b > d$ or $d > b)$ respectively. Use a uniform-cost development rule.

Try other selection rules including ones for which the decisions to be made are not functions of level only.

3.2 Modify the discussion of section 3.1 to relate directly to a maximisation problem (that is, without converting to an equivalent minimisation problem).

3.3 The arcs of a *directed tree* $(\mathcal{F}, T)$ are 'directed away from the root'. Prove that the number of nodes $|\mathcal{F}|$ of the tree satisfies $|\mathcal{F}| = 1 + \Sigma_{x \in \mathcal{F}} d_o(x)$, where $d_o(x)$ is the out-degree of x. Verify this for the trees of figure 3.4 all of which are rooted at the node corresponding to the set of all feasible solutions.

Hence or otherwise prove that a complete enumeration tree obtained by continual (proper) partitioning of a feasible set F must satisfy $|F| + 1 \leq |\mathcal{F}| \leq 2|\mathcal{F}| - 1$.

3.4 Solve the linear assignment problem of example 3.3
(a) using the lower bounding function g;
(b) using the lower bounding function g'.

Note that $g'(X) \geq g(X)$ for all $X \subset C$. Let P be a general problem with a pair g, g' of bounding functions satisfying this relation. Prove that the search tree generated using g' is contained in ('is a subtree of') the search tree generated using g, provided the same branching, selection and development rules are used.

3.5 (Multi-constraint knapsack problem – Shih, 1979) A supplier of Christmas hampers wishes to produce a standard package containing several different items of 'Christmas fare'. The choice is to be made from a selection of n items with item i having volume v_i, weight w_i and estimated potential profit of p_i. The hampers are of a fixed size and can accommodate a volume V of goods (this includes an allowance for packing problems due to shape). It is also decided that the maximum acceptable weight is W. The problem of maximising expected profit per hamper becomes

$$\text{maximise} \sum_{i=1}^{n} p_i x_i$$

subject to

$$\sum_{i=1}^{n} v_i x_i \leq V \qquad (1)$$

$$\sum_{i=1}^{n} w_i x_i \leq W \qquad (2)$$

x_i (the number of items of type i) is 0 or 1

This problem can be solved by B & B, as in example 3.5 with, for any node X, an upper bound $UB_v(X)$ [or $UB_w(X)$] being calculated by ignoring constraint (2) [constraint (1)]. Stronger bounds can be obtained by using the function (Shih, 1979)

$$UB(X) = \min\,[UB_v(X), UB_w(X)]$$

Solve this problem (see example 3.5) for the data set

i	1	2	3	4	
p_i	5	11	7	17	
v_i	2	4	3	5	$V = 8$
w_i	7	8	6	10	$W = 15$

3.6 (Flow shop sequencing)

(a) n jobs $J_1, \ldots, J_n$ are each to be processed on m machines $M_1, \ldots, M_m$ under the following conditions
  (i) all jobs are ready at time zero, at which time processing starts;
  (ii) J_i requires a processing time t_{ij} (known beforehand) on machine M_j, and processing, once started, must be completed without interruption;
  (iii) a machine can process only one job at a time;
  (iv) each job J_i is processed on the machines in the order $M_1, M_2, \ldots, M_m$;
  (v) the order of jobs is the same on each machine.

The (m, n) *flow shop sequencing* problem is to find a permutation $(i_1, \ldots, i_n)$ of $(1, 2, \ldots, n)$ so that the processing order $J_{i_1}, J_{i_2}, \ldots, J_{i_n}$ leads to minimal *makespan* (the time of completion of the last job on machine M_m). The chart in figure 3.10 shows the timing of jobs on machines given the following data and that the processing order J_1, J_2, J_3, J_4 is chosen.

	J_1	J_2	J_3	J_4
M_1	3	6	7	4
M_2	8	1	8	5
M_3	6	5	2	7

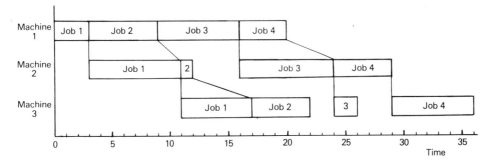

Figure 3.10

(b) Prove that the processing time $t(\pi)$ for any complete sequence π which starts with $\sigma = J_{i_1}, J_{i_2}, \ldots, J_{i_p}$ must satisfy (Ignall and Schrage, 1965)

$$t(\pi) \geq C(\sigma) + \sum_{i \notin \sigma} t_{i3} = g_3(\sigma)$$

where $C(\sigma)$ is the time at which job J_{i_p} is completed on M_m.

Find a minimal makespan sequence using the data of part (a) and a B & B method with a rank branching strategy and bounding function g_3.

Derive bounds $g_1(\sigma)$ and $g_2(\sigma)$ with respect to machines M_1 and M_2 and re-solve the problem using bounding function g defined by

$$g(\sigma) = \max\,[g_1(\sigma), g_2(\sigma), g_3(\sigma)]$$

If σ and σ' are two ordered sets containing the same subset of jobs and $g(\sigma) \leq g(\sigma')$ verify that σ dominates σ'.

3.7 During a production planning exercise, a range of n alternative products $P_1, \ldots, P_n$ are being considered where the production cost for P_i consists of a fixed cost k_i if any of P_i is produced and a variable cost of c_i per unit. The production level x_i of P_i is limited above by 2, that is, $x_i \leq 2$, and it is required that the total production level should reach a specified level corresponding to the constraint

$$\sum_{i=1}^{n} w_i x_i \geq W$$

A production policy is desired which minimises the total cost, subject to the conditions and constraints mentioned above. Formulate this problem as an MILP. Hence, or otherwise, show that the cost of any policy for which $x_i = 0$ if $i \in E$ and $x_i > 0$ if $i \in I$, is at least as high as the optimal value of the bounded variable knapsack problem

$$\text{minimise} \quad \sum_{i=1}^{n} c_i x_i + \sum_{i \in I} k_i$$

subject to

$$\sum_{i=1}^{n} w_i x_i \geq W \qquad \qquad (B)$$

$$0 \leq x_i \leq 2; \qquad x_i = 0 \text{ if } i \in E$$

Illustrate how the bound of (B) can be used in a B & B method for solving the fixed charge problem posed, by solving the particular problem for which $n = 4$, $W = 12$ and

i	1	2	3	4
c_i	3	4	1	2
k_i	4	4	2	1
w_i	4	3	1	1

(LU 1978, modified)

3.8 Solve the following 0–1 problem by the general (Balas) approach making full use of 'fix at zero' and 'fix at one' tests.

minimise $3x + 5y + 5z + 9t + 8u + 4v + 12w$

subject to

$$6x - 5y + 8z + t + 2u + 3v - w \leq 7$$

$$x + 2y - 9z - 3t - 2u - 2v - 2w \leq -10$$

$$x, y, z, t, u, v, w \in \{0, 1\}$$

4 Shortest Route Problems

Shortest routes (paths or chains) are of fundamental importance in network and combinatorial optimisation problems, both in their own right and as sub-problems of other optimisation problems. Shortest route problems have obvious relevance when a shortest (or quickest or least cost) route is required between two points of a transportation system. Similar problems arise when it is required to find a maximal flow or least cost flow augmenting chain (chapters 9 and 10) or a maximum reliability or maximum capacity route between two points. (For the latter two problems the objective function is not additive but of the forms given in exercise 2.7 parts (a) and (c) respectively.) Shortest paths between two vertices of a special network are sought in knapsack problems (section 2.4) and in the inspection and assembly line balancing problems discussed below (section 4.1 and exercise 4.2).

Shortest distances between all pairs of vertices of a network are needed in location problems (chapter 5) and a method for finding these is given in section 4.3. Finally there is the problem of finding not only the best route but also the second, third, ..., Kth best routes.

The problem of finding a longest path in a network arises in connection with project management (section 6.1) and is clearly equivalent to finding a corresponding shortest path in the network with arc distances negated, and so attention will be directed towards minimisation problems.

A shortest path problem is straightforward to solve if all arc lengths are positive. If some arc lengths are negative but there are no negative length circuits then the problem can still be solved satisfactorily (section 4.2) but the available methods are rather less efficient. If negative length circuits are present the problem is not sensible unless extra conditions are imposed. We shall assume throughout this chapter that, unless otherwise stated, no negative length circuits are present.

The vertices will be numbered 1 to n unless otherwise indicated and the following notation will be adopted

d_{ij} — the length of arc ij

$d(i,j)$ — the shortest distance from vertex i to vertex j

$d(j)$ — an abbreviation for $d(1,j)$

4.1 SHORTEST PATH BETWEEN TWO POINTS

Example 4.1 (White, 1969)   A production line consists of an ordered sequence of n production stages each of which consists of a manufacturing operation followed by a potential inspection station. The product being manufactured enters stage 1 in batches of size B. As each item of a batch moves along the production line faults may be created in it at any stage j with independent probability b_j. A fault created at stage j costs r_{jk} to repair if detected at stage k.

Inspection must be carried out at the nth stage but is optional at other stages. The cost of inspecting at stage k, if the batch was last inspected at stage j, is $f_{jk} + Bv_{jk}$, where f_{jk} is the fixed cost per batch and v_{jk} is the variable cost per item inspected. How should inspection effort be allocated so as to minimise total costs?

Solution   Suppose inspection occurs at stage j but not again until stage k. Then, if second order terms are ignored, on average $\sum_{\alpha=j+1}^{k} Bb_\alpha$ faults are created when stage k is reached, and have to be repaired at a total cost of $\sum_{\alpha=j+1}^{k} Bb_\alpha r_{\alpha k}$. Hence the total cost attributable to the 'gap' jk is

$$c_{jk} = f_{jk} + Bv_{jk} + B \sum_{\alpha=j+1}^{k} b_\alpha r_{\alpha k} \qquad 0 \leqslant j < k \leqslant n$$

The total cost incurred in producing a batch is thus

$$C = c_{0k_1} + c_{k_1 k_2} + \ldots + c_{k_{s-1} n}$$

if inspections occur at stages $k_1, k_2, \ldots, k_{s-1}$. The problem is to find $k_1, \ldots, k_{s-1}$ so as to minimise C. This is clearly the same problem as that of finding a shortest path from vertex 0 to vertex n in a network H on vertex set $\{0,1,2,\ldots,n\}$ and with arcs jk of length c_{jk} for all $0 \leqslant j < k \leqslant n$. Figure 4.1 shows the graph H for $n = 5$.

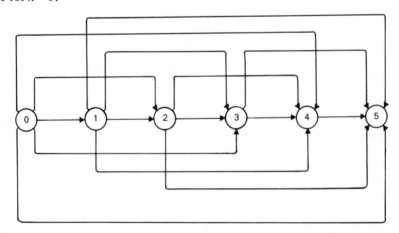

Figure 4.1 Model graph for the inspection problem (example 4.1) with $n = 5$

Shortest Route Problems

The above network has the special structure of having no circuits, a unique *source* (no incoming arcs) the vertex 0, and a unique *sink* (no outgoing arcs) the vertex n. (The same is true of the example networks of section 2.4.) Finding a shortest path from source to sink in such a network is a straightforward matter. The nodes of the search tree correspond to 'partial' paths of the form $0 i_1 i_2 i_3 \ldots i_p$. Any strategy generating the complete enumeration tree can be used with cutoffs occurring if either

(1) as short a path to i_p has already been found (principle of optimality); or
(2) a path from the source to the sink, whose length is as short, has already been found (bounding principle).

Moreover, since the network has no circuits, it does not matter if any, or even all, arc lengths are negative.

On the other hand, if a shortest path from b to c is sought where c is not the sink, the choice of search strategy can assume a greater importance. To see this consider the network of figure 3.4a in which it is required to find a shortest path from L to R.

Depth-first development, with ties being resolved by favouring vertices higher in the diagram, leads to the search tree of figure 3.4b whereas breadth-first development leads to the much smaller tree of figure 3.4c. The trouble, in this case, with the DF strategy is that the vertex aimed at is 'overshot'. This is remedied by the BF strategy which considers partial paths with a monotonically increasing number of arcs. However, it does not take sufficient account of bounds and the uniform-cost strategy reduces the size of the tree even further (figure 3.4d). This latter method is just that of Dijkstra (1959) and is set out below where, for convenience, it is assumed that the vertices are assigned distinct names $1, 2, \ldots, n$. $|p(x)|$ is the next vertex from x back along a shortest path from 1 to x. Γ denotes the successor mapping, an undirected graph being implicitly replaced by the corresponding symmetric directed graph.

Algorithm (D)
(Shortest path from 1 to n, all $d_{ij} \geq 0$)

Step 1 (Setup)
Set $d(1) = p(1) = 0$,
$\quad d(x) = p(x) = \infty \quad$ all $x > 1$,
$\quad u = 1$.

Step 2 (Revision of labels)
For all $x \in \Gamma u$:
if $[p(x) = \infty]$ or
$\quad [p(x) < 0$ and $d(u) + d_{ux} < d(x)]$
then set $p(x) = -u$ and $d(x) = d(u) + d_{ux}$.

Step 3 (Development)
    Choose u such that $d(u) = \min\limits_{p(x)<0} d(x)$.
    (If there are none then terminate; there is no path from 1 to n.)
    Replace $p(u)$ by $-p(u)$.
    If $u = n$ then terminate; otherwise go to step 2.

This algorithm should be compared with the one outlined in section 2.4. Search trees such as the one in figure 2.11 may contain several nodes x corresponding to a single vertex x of the network. However, it is clearly necessary to retain only one of these, and algorithm (D) does just this with the particular node x that is kept being specified by the pointer $p(x)$. Thus Dijkstra's algorithm not only grows a search tree but selectively prunes it as well. To see this consider the sequence of trees shown in figure 4.2 determined by the pointers $p(x)$ which are generated for the problem of finding the shortest path from S to F in the network of figure 2.10. The growth of the tree follows the pattern it did before until tree (3) is obtained. Node b having the least bound is now selected and node e added, and the pointer from c being switched from a to b as the length of Sbc is less than the length of Sac. The solution continues until tree (8) is reached. Node g is selected but $p(F)$ remains at h. F is the next node selected and so the algorithm terminates.

If $p(x) = \infty$ or $p(x) < 0$ then $d(x)$ is the *temporary label* of x; if $0 \leq p(x) < \infty$ then $d(x)$ is the *permanent label* of x. At any stage the set P of permanently labelled nodes together with arcs defined by the set of pointers $\{p(x)\}_{x \in P}$ forms a directed tree (P, Δ^{-1}). The corresponding reverse graph (that is, directions reversed) is a tree which will be termed a shortest distance or *SD*-tree on P for reasons made clear by the following results.

Theorem 4.1
When u is selected in step 3 of algorithm (D) $d(u)$ is equal to the distance from vertex 1 to vertex u.

Proof First note that for any x not on an SD-tree, the value of $d(x)$ is the length of a shortest path using only 'arcs of the tree' and one further arc. This is certainly true after the first application of step 2, and the condition is clearly maintained by the way in which labels are updated.

Now suppose that u is about to be added to the tree, that is $d(u) = \min\limits_{p(x)<0} d(x)$. If $d(u)$ is not equal to the distance from vertex 1 to vertex u then let π be a minimal length path from 1 to u. π must contain more than one 'non-tree' arc and hence vertices other than u which do not belong to the SD-tree. Let z be the first such vertex on π, let $\pi \equiv \pi_1 \pi_2$ where π_1 is the part of π from 1 to z and π_2 is the part from z to u, and denote the lengths of π, π_1 and π_2 by $l(\pi), l(\pi_1)$ and $l(\pi_2)$ respectively. Then $l(\pi_1) = d(z)$ since π is minimal and hence

$$d(z) + l(\pi_2) = l(\pi) < d(u) \tag{4.1}$$

Shortest Route Problems

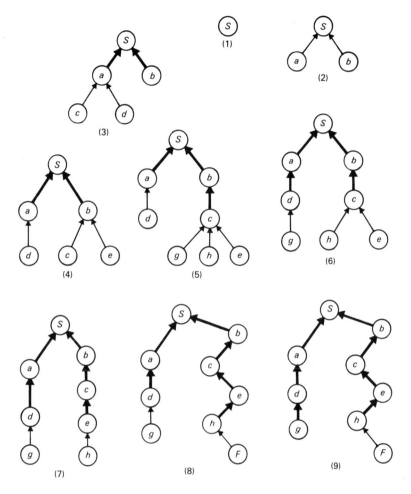

Figure 4.2 Trees generated by the application of algorithm (D) to the shortest distance problem shown in figure 2.10. Heavy and light arcs correspond to permanent and temporary labels respectively

Since all arc lengths are non-negative, $l(\pi_2) \geq 0$ and so

$$d(z) < d(u) \qquad (4.2)$$

But this contradicts the minimality of $d(u)$ and so $d(u)$ must be equal to the distance from vertex 1 to vertex u. □

Corollary If x is on an SD-tree then the unique path from 1 to x in the tree is also a shortest path from 1 to x in the original network and its length is $d(x)$.

Dijkstra's algorithm is very efficient, clearly being $O(n^2)$ for complete networks. (See section 4.2 for a discussion of computational experience.)

One way in which algorithm (D) might be improved is to reduce the number of nodes permanently labelled. First it is clear that finding a shortest path from 1 to n in a network G is equivalent to finding a shortest path from n to 1 in the reverse network G^{-1}, and the possibility of 'searching from both ends' suggests itself. That such a *bi-directional* search can lead to fewer labellings is seen by considering the symmetric network of figure 4.3 which for convenience is illustrated by an undirected graph.

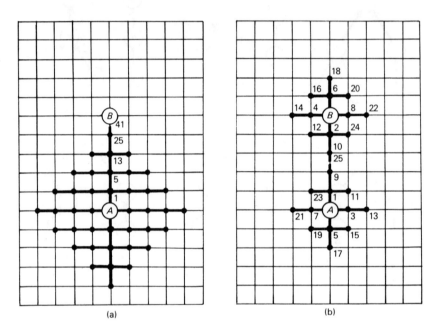

Figure 4.3 Search trees generated by (a) bi-directional, and (b) uni-directional versions of Dijkstra's algorithm

Lower and upper SD-trees are grown with nodes being added to the trees alternately. The order in which the nodes are added is shown by the numbers alongside (figure 4.3b). It is seen that node 25 on the lower tree coincides with node 10 of the upper tree and so a path π of length 5 has been found from A to B. Moreover, no shorter path from A to B can exist as any path other than π must contain two arcs from each of the lower and upper trees and at least one further arc.

For the corresponding *uni-directional* search (in which just one tree is grown, from A say) at least 41 nodes must be added to the tree (figure 4.3a). That is, the bi-directional search has led to a saving of nearly 40 per cent in the number of nodes permanently labelled. The precise numbers $N_U(D)$ and $N_B(D)$ of nodes added to the tree(s) in the uni-directional and bi-directional cases when A and B are distance D apart vertically will depend on how ties are broken but we can assert that for D odd

$$2D^2 - 2D + 1 \leqslant N_U(D) \leqslant 2D^2 + 2D$$
$$D^2 \leqslant N_B(D) \leqslant D^2 + 4D + 2$$

Thus

$$41 \leqslant N_U(5) \leqslant 60 \text{ and } 25 \leqslant N_B(5) \leqslant 47$$

and for $D \geqslant 7$, $N_U(D) > N_B(D)$ whatever the strategy for breaking ties. For D even, the comparison is more pronounced with $N_U(D) > N_B(D)$ for $D \geqslant 4$ whatever the tie-breaking strategy, provided nodes are again added alternately to the lower and upper trees for the bi-directional search.

The above analysis overlooks several factors.

(1) The saving in terms of nodes permanently labelled varies with the underlying network and the particular vertices between which a shortest path is sought.

(2) Four labels $d_A(x)$, $d_B(x)$, $p_A(x)$ and $p_B(x)$ would have to be set for all vertices x of the network at the outset of a bi-directional analogue of algorithm (D). For the above example this extra effort would outweigh the advantages from having smaller trees.

(3) The network of figure 4.3 is very special in that it is bipartite and all edges have the same length. Because of this the first complete path between A and B found by a bi-directional search will of necessity be a shortest path. This is not true in general though any shorter path cannot contain more than one arc not in either tree; why?

For these reasons it is usually not worth while performing a bi-directional search (but see section 9.2). However it can be advantageous if

(1) the whole network is not generated but only those vertices which become, or are adjacent to, a node on one of the trees, the overheads mentioned in (2) above thus being side-stepped;

(2) the first path found between A and B is acceptable as a solution even though it may fail to be optimal. This cuts out the extra bookkeeping involved in checking for termination.

This discussion should be sufficient to introduce bi-directionality and to suggest that it is potentially useful under the right conditions.

Another way in which the number of nodes permanently labelled may be reduced is, in the spirit of B & B, to take into account not only the distance $d(x)$ from A to x but also some estimate of the further distance from x to B.

Definition 4.1
A function h on the vertices of a network (X, Γ, d) is *consistent* if for each arc xy

$$h(x) \leq h(y) + d_{xy}$$

Theorem 4.2
For a consistent function h on a network (X, Γ, d)
(1) if $h(B) = 0$ then for all $x \in X$, $h(x)$ is a lower bound on the distance from x to B;

(2) for any path $u_1 u_2 \ldots u_s$,

$$h(u_1) \leq h(u_s) + (d_{u_1 u_2} + \ldots + d_{u_{s-1} u_s}).$$

Proof Direct from the definition of consistency.

If a consistent function is available the first result of the above theorem together with the discussion leading up to algorithm (D) suggests the following algorithm due to Nemhauser (1972) (and Hart *et al.*, 1968).

Algorithm (N)
(To find the shortest path from 1 to n)

Step 1 (Setup)
Set $d(1) = p(1) = 0$,
$\quad d(x) = p(x) = \infty \quad$ all $x > 1$,
$\quad u = 1$.

Step 2 (Revision of labels)
For all $x \in \Gamma u$
if $[p(x) = \infty]$ or
$\quad [p(x) < 0 \text{ and } d(u) + d_{ux} < d(x)]$
then set $p(x) = -u$ and $d(x) = d(u) + d_{ux}$.

Step 3 (Development)
Choose u such that $f(u) = \min_{p(x) < 0} f(x)$
where $f(x) = d(x) + h(x)$.
(If no such u exists then terminate; there is no path from 1 to n.)
Replace $p(u)$ by $-p(u)$.
If $u = n$ then terminate; otherwise go to step 2.

The only difference from algorithm (D) is in the use of f in place of d for selecting the next node to be added to the tree which we will again call an SD-tree (see corollary to theorem 4.3 below for justification).

Theorem 4.3
When u is selected in step 3 of algorithm (N) $d(u)$ is equal to the distance from vertex 1 to vertex u provided h is a consistent function.

Proof The proof is identical to that for theorem 4.1 up to inequality 4.1. Then adding $h(u)$ to both sides gives

$$d(u) + h(u) > d(z) + l(\pi_2) + h(u)$$
$$\geqslant d(z) + h(z) \quad \text{by theorem 4.2b}$$

but this contradicts the minimality of $d(u) + h(u)$. Hence $d(u)$ is equal to the distance from vertex 1 to vertex u. □

Corollary 1 If x is on an SD-tree obtained using algorithm (N) then the unique path from 1 to x in the tree is a shortest path from 1 to x in the original network and its length is $d(x)$.

Corollary 2 Nodes are added to the tree in order of increasing $d(x) + h(x)$.

Example 4.2 Find a consistent function h, other than the trivial $h = 0$, for the network of figure 4.4. Use this h and algorithm (N) to find the shortest path from A to B.

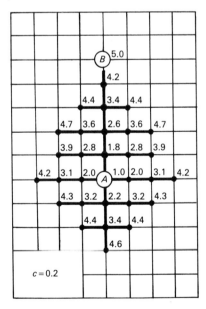
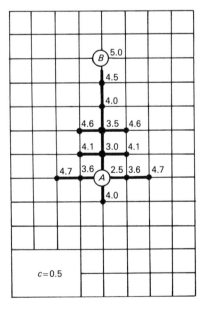

Figure 4.4 Two examples of the application of algorithm (N) with $c = 0.2$ and $c = 0.5$ (see example 4.2)

Solution Choosing the 'bottom and left edges' of the network as coordinate axes the vertices can be assigned integer coordinates (α, β) with $0 \leqslant \alpha \leqslant 8$ and $0 \leqslant \beta \leqslant 12$. The true distance between vertices x and y sited at (α_x, β_x) and (α_y, β_y) respectively, is clearly

$$d(x,y) = |\alpha_x - \alpha_y| + |\beta_x - \beta_y|$$

The 'crow-fly' distance $h(x, y) = \{(\alpha_x - \alpha_y)^2 + (\beta_x - \beta_y)^2\}^{1/2}$ satisfies $0 \leqslant h(x, y) \leqslant d(x, y)$

Define $h(x) = h(x, n)$ for all vertices x then for any pair of adjacent vertices i and j

$$\begin{aligned} h(i) &= h(i, n) \leqslant h(j, n) + h(i, j) \quad \text{(triangle property)} \\ &= h(j) + h(i, j) \\ &= h(j) + d_{ij} \quad \text{(i and j are adjacent)} \end{aligned}$$

That is, h is a consistent function for the given network. The application of algorithm (N) using this h leads directly to the minimal path with five nodes being added to the SD-tree. If A and B are distance 5 apart but AB is not parallel to one of the axes then a few more nodes are added to the tree.

For this example ch is also a consistent function and the SD-trees are shown in figure 4.4 for $c = 0.2$ and 0.5. If N denotes the number of nodes added to the tree then

c	0	0.2	0.5	1
N	41–60	30–32	14	5

Notice that even a small value of c leads to a substantial reduction in the number of nodes permanently labelled.

With careful programming $h(x)$ need be calculated only once for each x on the tree or adjacent to a node on the tree. It can thus be seen that the use of algorithm (N) can be advantageous provided the calculation of $h(x)$ is not 'too difficult' compared to the effort of adding an extra node to the tree. Also algorithm (N) is valid when negative length arcs are present though algorithm (D) is inapplicable.

There is still the question of finding a suitable h and establishing consistency. This may not be easy, particularly if negative length arcs are present, even though it is readily proved that a consistent function does exist.

4.2 THE SHORTEST PATH PROBLEM: GENERAL CASE

In the previous section methods were presented for finding the shortest path between two specified vertices in a network with no negative length arcs. The

Shortest Route Problems

more general case can be solved by algorithm (N) but the problem of finding a consistent function may preclude its use.

It is also true that algorithm (D) can be extended to the general case by omitting the condition '$p(x) < 0$' from step 2; that is replacing step 2 by step 2'.

Step 2' (Revision of labels)

For all $x \in \Gamma u$
if $p(x) = \infty$ or
$$d(u) + d_{ux} < d(x)$$
then set $p(x) = -u$ and $d(x) = d(u) + d_{ux}$.

The algorithm as modified is $O(n^2)$ if all arcs have positive length but Johnson (1973) has shown that there exists a class of networks which require $O(n2^n)$ elementary operations. Despite this it might be worth while using this method if only a few negative length arcs are present and it is known in advance that no negative length circuits exist.

An $O(n^3)$ algorithm attributed to Ford (1956) (and also to Moore and Bellman), which is applicable in the general case and which can detect the presence of negative length circuits will now be described.

Assuming the vertices to be numbered 1 to n as before then, by the principle of optimality

$$d(1,1) = 0 \tag{4.3a}$$

$$d(1,j) = \min_k [d(1,k) + d_{kj}] \tag{4.3b}$$

Approximations $d^{(r)}(i)$ to $d(1,i)$, which clearly improve monotonically as r increases are obtained by setting

$$d^{(1)}(1) = 0$$
$$d^{(1)}(j) = d_{1j} \qquad j = 2,3,\ldots,n$$

and repeatedly using the equations

$$d^{(r+1)}(j) = \min \{ d^{(r)}(j), \min_{k \neq 1, j} [d^{(r)}(k) + d_{kj}]\}$$
$$= \min_{k \neq 1} [d^{(r)}(k) + d_{kj}] \qquad \text{if } d_{jj} = 0 \text{ all } j$$

Since the sequence $d^{(r)}(j)$ is bounded below [by $d(j)$] we might expect it to converge. The following result shows that it converges to the desired limit.

Theorem 4.4

For any vertex j which has a shortest path from 1 to j that contains r or fewer arcs, $d^{(r)}(j) = d(j)$.

Proof This result is readily established by mathmatical induction. A proof is not given here as the result follows from the more general one of theorem 4.5.

Corollary 1   If there are no negative length circuits present then $d^{(n-1)}(j) = d(j)$ all j.

Proof   Since all circuits have non-negative length the pointers $p(i)$ will at all times determine a tree and hence elementary shortest paths at termination. Since an elementary path cannot have more than $n-1$ arcs the result follows.

Corollary 2   The outlined algorithm, to be given formally as algorithm (Fd), is $O(n^3)$.

Corollary 3   If $d^{(p+1)}(j) = d^{(p)}(j)$ for all j and $p < n-2$ then $d^{(p)}(j) = d(j)$ all j.

Corollary 4   If the possibility of negative length circuits is permitted then the existence of such a circuit can be detected by checking whether $d^{(n)}(j) = d^{(n-1)}(j)$ all j.

We are now in a position to give a formal statement of Ford's algorithm in which $d_{ij} = \infty$ is assumed if there is no arc from i to j.

Algorithm (Fd)

Step 1 (Setup)
　　Set $d^{(1)}(1) = 0$, $p(1) = 0$ and $r = 1$.
　　Set $d^{(1)}(j) = d_{1j}$ and $p(j) = 1$　all $j \neq 1$.

Step 2 (Update)
　　Set change ← *false*.
　　For all $j \neq 1$ perform step 2a.

　　Step 2a
　　　Set $a = d^{(r)}(k^*) + d_{k^**j} = \min_{k \neq 1} [d^{(r)}(k) + d_{kj}]$.
　　　If $a < d^{(r)}(j)$
　　　then set change ← *true*, $d^{(r+1)}(j) = a$, $p(j) \leftarrow k^*$
　　　else set $d^{(r+1)}(j) = d^{(r)}(j)$.

Step 3 (Termination)
　　If change = *false* terminate.
　　If $r = n-2$ then terminate;
　　else replace r by $r+1$ and return to step 2.

It should be noted that for other than very dense networks efficiency is improved by restricting the minimisation of step 2a to the set $\Gamma^{-1}j$.

Shortest Route Problems 81

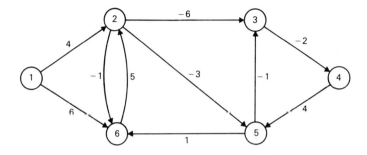

Figure 4.5

Example 4.3 Use algorithm (Fd) to find the shortest path from vertex 1 to vertex 6 in the network in figure 4.5.

Solution The details of the calculations are as follows

$d^{(1)}(1) = 0, \ d^{(1)}(2) = 4, \ d^{(1)}(3) = d^{(1)}(4) = d^{(1)}(5) = \infty, \ d^{(1)}(6) = 6$
$p(1) = 0, \ p(2) = p(3) = p(4) = p(5) = p(6) = 1$

$d^{(2)}(2) = \min[4, \min(6+5)] = 4$
$d^{(2)}(3) = \min[\infty, \min(4-6, \infty-1)] = -2 \qquad p(3) = 2$
$d^{(2)}(4) = \min[\infty, \min(\infty-2)] = \infty$
$d^{(2)}(5) = \min[\infty, \min(4-3, \infty+4)] = 1 \qquad p(5) = 2$
$d^{(2)}(6) = \min[6, \min(4-1, \infty+1)] = 3 \qquad p(6) = 2$

$d^{(3)}(2) = \min[4, \min(3+5)] = 4$
$d^{(3)}(3) = \min[-2, \min(4-6, 1-1)] = -2$
$d^{(3)}(4) = \min[\infty, \min(-2-2)] = -4 \qquad p(4) = 3$
$d^{(3)}(5) = \min[1, \min(4-3, \infty+4)] = 1$
$d^{(3)}(6) = \min[3, \min(4-1, 1+1)] = 2 \qquad p(6) = 5$

$d^{(4)}(2) = \min[4, \min(2+5)] = 4$
$d^{(4)}(3) = \min[-2, \min(4-6, 1-1)] = -2$
$d^{(4)}(4) = \min[-4, \min(-2-2)] = -4$
$d^{(4)}(5) = \min[1, \min(4-3, -4+4)] = 0 \qquad p(5) = 4$
$d^{(4)}(6) = \min[2, \min(4-1, 1+1)] = 2$

Although there was no change in the label of vertex 6, $d^{(4)}(5) < d^{(3)}(5)$ and so *change* will be *true*, and the algorithm does not terminate.

$d^{(5)}(6) = \min[2, \min(4-1, 0+1)] = 1 \qquad p(6) = 5$

Since $r = n - 2$ the algorithm can be terminated if it is known that no negative length circuits exist. Otherwise it is necessary to calculate $d^{(5)}(2), d^{(5)}(3), d^{(5)}(4)$ and $d^{(5)}(5)$ and then perform one more iteration.

It will be observed that at any iteration $r \geq 1$ not all of the labels $d^{(r)}(j)$ will be altered and the proportion needing to be updated may be very small particularly in the latter stages of the solution of larger problems. This suggests that time may be saved if the labels $d^{(r)}(j)$ which will need updating can be anticipated and leads on to the following modification of algorithm (Fd).

Algorithm (Fd')

Step 1 (Setup)
    Set $d'(1) = 0$, $p(1) = 0$.
    Set $d'(j) = d_{1j}$ and $p(j) = 1$ all $j \neq 1$.
    Set change = *true*.

Repeat step 2 while change is *true*.

Step 2 (Update)
    Find an arc kj for which

$$a = d'(k) + d_{kj} < d'(j).$$

    If there are none such, then set change ← *false*;
    else set $d'(j) \leftarrow d'(k) + d_{kj}$ and $p(j) \leftarrow k$.

Step 3 (Termination)
    The algorithm terminates with $d'(j) = d(j)$ for all j,
    and with an SD-tree being determined by the pointers $p(j)$.

Theorem 4.5
Provided there are no negative length circuits present algorithm (Fd') terminates with the correct distances $d(j)$ and a valid SD-tree.

Proof   Since there are no negative length circuits the pointers will always determine a tree and hence an elementary path from 1 to j for all j. Now, there are only a finite number of elementary paths from 1 to j and so $d'(j)$ [and hence $p(j)$] can only change a finite number of times. Therefore the algorithm will terminate after a finite number of steps.

Suppose now that the algorithm terminates with $d'(j) > d(j)$ for some vertex j. Let π^* be an elementary shortest path from 1 to j. Since $d'(1) = d(1)$ and $d'(i) \geq d(i)$ for all i on π^* it follows that there must be successive vertices r and s on π^* for which $d'(r) = d(r)$ and $d'(s) > d(s)$. Then

$$d'(s) > d(s) = d(r) + d_{rs} = d'(r) + d_{rs}$$

but this is impossible as it would imply that $d'(s)$ could be updated and so the algorithm could not have terminated! Thus the algorithm terminates with the correct distances. That the pointers provide an SD-tree follows directly. □

Algorithms of the form of (Fd') are called *label-correcting* algorithms as opposed to *label-setting* algorithms which are based on algorithm (D). How are arcs leading to a reduction of some vertex label to be selected effectively?

Suppose a first such arc ij has been found and $d'(j)$ can be updated by an amount e. It is easily seen that so can the label at each other vertex in the *subtree at j* (that is, the subtree on the set of vertices k such that the pointers lead from k to j). Dial et al. (1977) implemented this idea and found that it led to a substantial reduction in computing time. They were able to solve, within 1 second of CDC 6600 time on average

(1) randomly connected networks with 1000 vertices and up to 30 000 random length arcs;
(2) grid-based networks with 2500 vertices and around 9000 random length arcs.

Indeed, on the basis of these results it was found that their label-correcting algorithm performed better than their label-setting algorithm for sparse networks, but that label-setting may have the edge for denser networks. Golden (1976), using a different selection rule to test for updating of labels, found that his label-correcting algorithm was about twice as fast as his label-setting algorithm for a set of sparse networks. (These networks were on 50–1000 randomly placed vertices with random connections between vertices whose out-degrees were constrained to be in the range 2 to 6. Lengths of arcs were Euclidean distances.) However, Denardo and Fox (1979) have argued in favour of label-setting algorithms if implemented using appropriate computer science techniques. Their results suggest that these refinements might lead to faster times for very large networks. The present situation seems to be that for sparse networks with positive arc lengths there is probably not much to choose between sophisticated implementations of the two types of algorithm. For dense networks label-setting is to be preferred. If there is the possibility of negative length arcs then label-correction should be used.

Example 4.4 (Dantzig et al., 1967)   A large company moves regular amounts of high value goods between various sites. At present carriage of b_{ij} units, per unit time, from site i to site j is contracted to an outside concern at a cost of v_{ij} per unit. A light aircraft becomes available, at a hiring cost c per unit time and could be used for carrying some of the goods. Suppose the aircraft can carry w_{ij} units from site i to site j in time t_{ij} (including loading, etc.) at a cost of p_{ij} per unit. How might the decision of whether or not to use the aircraft be made?

Solution   Suppose the aircraft flies around a circuit $\pi = ijk \ldots mi$ then in time $T_\pi = \sum_\pi t_{\alpha\beta}$, it can carry $\sum_\pi \xi_{\alpha\beta}$ units at a total cost of $T_\pi c + \sum_\pi p_{\alpha\beta} \xi_{\alpha\beta}$, where

$$\xi_{\alpha\beta} = \min(b_{\alpha\beta}, w_{\alpha\beta})$$

and $\sum_{\pi}$ implies summation over arcs in π only.

Thus it is worth while using the aircraft if there is a circuit π for which

$$T_\pi c + \sum_\pi p_{\alpha\beta}\xi_{\alpha\beta} < \sum_\pi v_{\alpha\beta}\xi_{\alpha\beta} \tag{4.4}$$

Consider now the complete network G on the set of sites concerned, with the length $d_{\alpha\beta}$ of an arc $\alpha\beta$ defined by

$$d_{\alpha\beta} = ct_{\alpha\beta} + (p_{\alpha\beta} - v_{\alpha\beta})$$

The length of any circuit π is

$$c \sum_\pi t_{\alpha\beta} + \sum_\pi (p_{\alpha\beta} - v_{\alpha\beta})\xi_{\alpha\beta}$$

and from (4.4) it is seen that using the aircraft is definitely worth while if a negative length circuit can be found in the network G with arc lengths given by the matrix (d_{ij}). This can be checked by using algorithm (Fd).

Of course if a negative length circuit does exist we would like the one that leads to the largest saving per unit time. To do this we use the modified matrix $(d_{ij}(\lambda))$ where $d_{ij}(\lambda) = d_{ij} + \lambda t_{ij}$. Values λ_l and λ_u are found such that $\lambda = \lambda_l$ leads to negative length circuits and $\lambda = \lambda_u$ leads only to positive length circuits. A binary search procedure can be used to find, to within a given tolerance, a value λ_0 of λ which leads to a zero length circuit but no negative length ones. This circuit is then a most profitable one to use. Note that there is the possibility of further savings if goods can be carried from i to j when there is an intermediate call, at k say.

Christofides (1975) explains how algorithm (Fd) can be improved if it is being used specifically to detect negative length circuits.

4.3 OTHER SHORTEST PATH PROBLEMS

Example 4.5 A sports centre is planned for the use of several separate communities j with populations $w_j, j = 1, \ldots, n$. It is decided to site the centre in one of the communities and so that the average distance travelled is minimised. How should the decision be made?

Solution For the moment we make the simple but unrealistic assumption that a fixed proportion α of each community will use the centre whichever site is chosen. (A more realistic situation is discussed in example 5.5 section 5.2.)

If the centre is at i then the average distance travelled (there and back) is

$$\frac{\sum_{j=1}^{n} 2\alpha w_j d(i,j)}{\sum_{j=1}^{n} \alpha w_j}$$

Shortest Route Problems

The problem thus becomes that of calculating

$$\tau(i) = \sum_{j=1}^{n} w_j d(i,j) \quad \text{for all } i$$

and selecting the site for which $\tau(i)$ is minimal.

Clearly for this and many other such location problems it is necessary to know all (or at least many) of the intervertex distances $d(i, j), i, j = 1, \ldots, n$. An elegant and simple algorithm for achieving this, due to Floyd (1962), follows.

Algorithm (Fl)
(To find all intervertex distances)

Step 1 (Setup)
  Set $d^{(0)}(i, j) = d_{ij}$ all i, j.
  $\theta(i, j) = i$.

For $k = 1, 2, \ldots, n$ perform step 2.

Step 2 (Iteration)
  For all $i \neq k$ with $d^{(k-1)}(i, k) \neq \infty$
  For all $j \neq i, k$ with $d^{(k-1)}(k, j) \neq \infty$
    If $a = d^{(k-1)}(i, k) + d^{(k-1)}(k, j) < d^{(k-1)}(i, j)$
    then set $d^{(k)}(i, j) = a$
      and $\theta(i, j) \leftarrow \theta(k, j)$
    else set $d^{(k)}(i, j) = d^{(k-1)}(i, j)$

Example 4.6 Use Floyd's algorithm to find the shortest distances (and paths) between all pairs of vertices for the network in figure 4.6.

Solution Denoting $[d^{(k)}(i, j)]$ and $[\theta(i, j)]$ by $D^{(k)}$ and Θ respectively

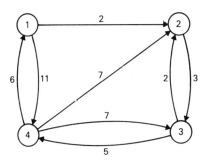

Figure 4.6

$$D^{(0)} = \begin{bmatrix} 0 & 2 & \infty & 11 \\ \infty & 0 & 3 & \infty \\ \infty & 2 & 0 & 5 \\ 6 & 7 & 7 & 0 \end{bmatrix} \quad \Theta \leftarrow \begin{bmatrix} 1 & 1 & 1 & 1 \\ 2 & 2 & 2 & 2 \\ 3 & 3 & 3 & 3 \\ 4 & 4 & 4 & 4 \end{bmatrix}$$

$$D^{(1)} = \begin{bmatrix} 0 & 2 & \infty & 11 \\ \infty & 0 & 3 & \infty \\ \infty & 2 & 0 & 5 \\ 6 & 7 & 7 & 0 \end{bmatrix} \quad \Theta \leftarrow \begin{bmatrix} 1 & 1 & 1 & 1 \\ 2 & 2 & 2 & 2 \\ 3 & 3 & 3 & 3 \\ 4 & 4 & 4 & 4 \end{bmatrix}$$

$$D^{(2)} = \begin{bmatrix} 0 & 2 & 5 & 11 \\ \infty & 0 & 3 & \infty \\ \infty & 2 & 0 & 5 \\ 6 & 7 & 7 & 0 \end{bmatrix} \quad \Theta \leftarrow \begin{bmatrix} 1 & 1 & 2 & 1 \\ 2 & 2 & 2 & 2 \\ 3 & 3 & 3 & 3 \\ 4 & 4 & 4 & 4 \end{bmatrix}$$

$$D^{(3)} = \begin{bmatrix} 0 & 2 & 5 & 10 \\ \infty & 0 & 3 & 8 \\ \infty & 2 & 0 & 5 \\ 6 & 7 & 7 & 0 \end{bmatrix} \quad \Theta \leftarrow \begin{bmatrix} 1 & 1 & 2 & 3 \\ 2 & 2 & 2 & 3 \\ 3 & 3 & 3 & 3 \\ 4 & 4 & 4 & 4 \end{bmatrix}$$

$$D^{(4)} = \begin{bmatrix} 0 & 2 & 5 & 10 \\ 14 & 0 & 3 & 8 \\ 11 & 2 & 0 & 5 \\ 6 & 7 & 7 & 0 \end{bmatrix} \quad \Theta \leftarrow \begin{bmatrix} 1 & 1 & 2 & 3 \\ 4 & 2 & 2 & 3 \\ 4 & 3 & 3 & 3 \\ 4 & 4 & 4 & 4 \end{bmatrix}$$

As an example of the working consider the calculation of $d(1, 4)$. $d^{(2)}(1, 4) = d^{(1)}(1, 4) = d^{(0)}(1, 4) = 11$. At the $k = 3$ iteration with $i = 1, j = 4, i \neq k \neq j \neq i$ and

$$d^{(2)}(1, 3) \neq \infty, \quad d^{(2)}(3, 4) \neq \infty$$

and so we compare

$$d^{(2)}(1, 4) = 11 \text{ with } d^{(2)}(1, 3) + d^{(2)}(3, 4) = 10$$

Since the latter is smaller, $d^{(3)}(1, 4)$ becomes 10 and $\theta(1, 4)$ is replaced by $\theta(3, 4)$ which at this stage is 3. $d(1, 4) = d^{(4)}(1, 4) = d^{(3)}(1, 4) = 10$. The shortest path from 1 to 4 is obtained from Θ.

$\theta(1, 4) = 3$ that is 34 is the last arc on π
$\theta(1, 3) = 2$ that is 23 is the next last arc on π
$\theta(1, 2) = 1$ that is 12 is the first arc on π

Thus 1234 is the shortest path from vertex 1 to vertex 4 and its length is 10.

Theorem 4.6
At termination, Floyd's algorithm yields
(1) $d(i, j) = d^{(n)}(i, j)$;
(2) the last arc on a shortest path from i to j is $\theta(i, j)$.

Proof (1) The proof is by induction using as hypothesis $H(k) : d^{(k)}(i, j)$ is the shortest distance from i to j using only vertices $1, 2, \ldots, k$ as possible intermediate vertices.

If π contains k as an intermediate vertex let π_1, π_2 be the parts of π from i to k and k to j respectively. Then π_1 must be a shortest path from i to k and π_2 a shortest path from k to j each using only $1, 2, \ldots, k-1$ as possible intermediate vertices. Hence by $H(k-1)$

$$d^{(k)}(i, j) = d^{(k-1)}(i, k) + d^{(k-1)}(k, j)$$

as required by $H(k)$. If π does not contain k then $d^{(k)}(i, j) = d^{(k-1)}(i, j)$ and $H(k)$ is again satisfied. Thus the truth of $H(k-1)$ implies the truth of $H(k)$ and since $H(1)$ is true, result (1) follows.

Result (2) is seen to be true by a similar argument. □

Algorithm (Fl) has been improved by Yen (1972). Spira (1973) presented an algorithm which took $O(n^2 \log \log n)$ steps on average (though the worst case still requires $O(n^3)$ steps). Another scheme for finding shortest distances (and paths) between all pairs of vertices is given by Dantzig (1967).

Sometimes it is not sufficient to know just a shortest path from i to j, the 2nd, 3rd, $\ldots$, Kth shortest paths also being required in case the shortest is unsatisfactory in some way. This is so if it is required that the shortest path satisfying some further constraint(s) be found where the extra constraint(s) are difficult to incorporate or imprecisely defined. In the first case the 1st, 2nd, 3rd, $\ldots$ best paths are generated until one is found which also satisfies the extra constraint(s). In the second case the first K (predetermined) best paths may be generated and presented to the decision maker for him to choose from in the light of the extra constraint(s). Most problems involving paths other than the 1st best are only concerned with elementary paths and this restriction will be imposed here. If paths incorporating circuits are also of interest then the algorithm of Dreyfus (1969) may be used.

We first consider the 2nd best path. It should be noted that this may in fact have the same length as the designated 1st best path if there are ties. Let π be a path from 1 to n, and i a vertex on π. Then a *deviation* $\pi(i, \alpha)$ of π at i, where

α is *not* the vertex after i on π, is a path which coincides with π from 1 to i then goes directly to α then by a shortest path to n. The part of π up to i, denoted by $\text{rt}(\pi, i)$ is the *root* of the deviation, and the remaining part of $\pi(i, \alpha)$ to n, denoted by $\text{sp}(\pi, i, \alpha)$, is the *spur* of the deviation.

Theorem 4.7
The 2nd best path from 1 to n is a deviation $\pi^1(i, \alpha)$ of the 1st best path π^1.

Proof Let π^2 be the path designated 2nd best, and assume that π^2 coincides with π^1 up to some vertex i (which may be 1) but then deviates to some other vertex α. Then if π^2 gives a shortest path from α to n it is a deviation $\pi^1(i, \alpha)$. Otherwise it is inferior to $\pi^1(i, \alpha)$ and so cannot be 2nd best. □

Corollary The 2nd best elementary path from 1 to n is a deviation $\pi^1(i, \alpha)$ with α not belonging to $\text{rt}(\pi^1, i)$. Moreover, the length of $\pi^1(i, \alpha)$ is not greater than the length of $\pi^1(i, \beta)$ for any $\beta \notin \text{rt}(\pi^1, i)$.

The 2nd best elementary path from 1 to n can be obtained by first finding for each j, a shortest path P_j^1 from the set $Q_j(\pi^1)$ of proper deviations at j. A shortest path of $\{P_j^1\}$, P_i^1 say, is then taken as the 2nd best elementary path from 1 to n.

The 3rd best elementary path from 1 to n is a deviation of either the 1st or 2nd best (or both). That is, it is one of the set $\{P_j^1\}_{j \neq i}$ or a proper deviation of π^2 at m for some $m \in \text{sp}(\pi^1, i, \alpha)$. Thus it is only necessary to generate the sets $Q_m(\pi^2)$ where

$$Q_m(\pi^2) = \{\text{deviations } \pi^2(m, \beta) \mid m \in \text{sp}(\pi^1, i, \alpha)\}, \quad m \neq i$$

$$Q_i(\pi^2) = \{\text{deviations } \pi^2(i, \beta) \mid \beta \neq \text{ successor of } i \text{ in } \pi^1\}$$

The 4th, 5th, ... best paths are found in a similar way. The algorithm, due to Yen (1971) can now be described.

Algorithm (Y)
(To find the K shortest paths)

Step 1 (Setup)
 Find a shortest path π^1 from 1 to n using algorithm (D) or algorithm (Fd) as appropriate.
 Set LIST = ϕ.
 Set $k = 1$.

Step 2 (Augmentation)
 If $k = 1$ then set $i = 1$;
 else suppose $\pi^k = \pi^q(i, \alpha)$ for some $1 \leq q < k$.
 For all $m \in \text{sp}(\pi^k, i, \alpha)$ find a shortest elementary path P_m^k in

$Q_m(\pi^k) = \{\pi^k(m, \beta) \mid \text{with } d_{m\gamma} \leftarrow \infty \text{ (temporarily) if there is a } j < k \text{ such that } \pi^k = \pi^j(m, \gamma)\}$.

Adjoin P_m^k to LIST.

Step 3 [Designation of $(k+1)$th best path]

If LIST $= \phi$ then terminate with the best k paths; no more exist.
Remove a shortest element of LIST and designate it π^{k+1}.
Replace k by $k+1$.
If $k = K$ then terminate; otherwise go to step 2.
Since only the best K paths are desired the size of LIST need never exceed $K - k$ elements.

Example 4.7 Find the best 6 paths from 1 to 7 in the network of figure 4.7.

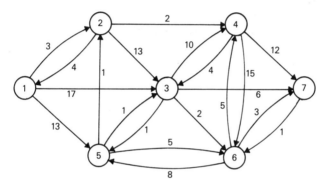

Figure 4.7

Solution
$\pi^1 = 12437$ (length 15) is a shortest path.
LIST $\leftarrow \phi$.
Set $d_{12} \leftarrow \infty$ to get
  $P_1^1 = 1537$ (length 20) as best deviation of π^1 at 1.
Set $d_{24} \leftarrow \infty$ to get
  $P_2^1 = 1237$ (length 22) as best deviation of π^1 at 2.
Set $d_{43} \leftarrow \infty$ to get
  $P_4^1 = 1247$ (length 17) as best deviation of π^1 at 4.
Set $d_{37} \leftarrow \infty$ to get
  $P_3^1 = 1243567$ (length 18) as best deviation of π^1 at 3.
(It is true that 1243537 (length 17) is shorter than P_3^1 but it contains a repeated vertex and so is not admissible.)
LIST $\leftarrow \{P_1^1, P_2^1, P_4^1, P_3^1\}$.
The shortest element of LIST is P_4^1 and so
$\pi^2 = 1247$.

LIST ← $\{P_1^1, P_2^1, P_3^1\}$.
Set $d_{47} \leftarrow \infty$ to get
 $P_4^2 = 12467$ (length 27) as best deviation of π^2 at 4.
LIST ← $\{P_1^1, P_2^1, P_3^1, P_4^2\}$.
$\pi^3 = 1243567$.
LIST ← $\{P_1^1, P_2^1, P_4^2\}$.
Since π^1 and π^2 both start with 1243 both d_{37} and d_{35} must be set to ∞ to get
 $P_3^3 = 124367$ (length 21) as best deviation of π^3 at 3.
(1243537 is again not admissible.)
LIST ← $\{P_1^1, P_2^1, P_4^2, P_3^3\}$ or since only the best 6 paths are required the longest element P_4^2 can be removed.
LIST ← $\{P_1^1, P_2^1, P_3^3\}$.
$\pi^4 = 1537$.
LIST ← $\{P_2^1, P_3^3\}$.
Set $d_{12} \leftarrow d_{15} \leftarrow \infty$ to get
 $P_1^4 = 137$ (length 23) as best deviation of π^4 at 1.
Set $d_{53} \leftarrow \infty$ to get
 $P_2^4 = 1567$ (length 21) as best deviation of π^4 at 5.
Set $d_{37} \leftarrow \infty$ to get
 $P_3^4 = 15367$ (length 27) as best deviation of π^4 at 3.
LIST ← $\{P_2^1, P_3^3, P_1^4, P_2^4, P_3^4\}$ or since only two more paths are required.
LIST ← $\{P_3^3, P_2^4\}$.
$\pi^5 = 124367$ (P_3^3).
Since P_2^4 is also of length 21 we can immediately set $\pi^6 = 1567$ (P_2^4).

The most computationally expensive part of the algorithm clearly arises from the many shortest path problems that have to be solved in step 2. If negative length arcs are present then the use of algorithm (Fd) to find the P_m^k leads to an $O(Kn^4)$ algorithm. However, the following result allows us to reduce this to $O(Kn^3)$.

Theorem 4.8
If π is a shortest path from 1 to n in a network (X, Γ, d) then it is also a shortest path from 1 to n in (X, Γ, d') where
$$d'_{ij} = d_{ij} + u_i - u_j$$
for any set of numbers $u_i, i = 1, 2, \ldots, n$.

Proof The length of every path from 1 to n is changed by the same amount $u_1 - u_n$ and so a shortest path remains a shortest path. □

Corollary Choosing $u_i = d(1, i)$ leads to a network with no negative length arcs.

Shortest Route Problems

Proof

$$d'_{ij} = d_{ij} + d(1, i) - d(1, j)$$
$$\geq [d(1, i) + d(i, j)] - d(1, j)$$
$$\geq 0 \text{ by the triangle property.} \square$$

The transformation of the above theorem cannot introduce any negative length circuits, since the length of every circuit is non-negative. Hence algorithm (D) is applicable to the modified network for the particular choice of u_i. The complexity of Yen's algorithm is thus improved to $O(Kn^3)$. Of course a label-correcting algorithm may still be required in step 1.

Horne (1980) reports an application to a network with 152 vertices, 240 arcs and no circuits, which arose in connection with minimising dairy management costs. No computing times were given but he stated that for $n > 7$ he found for this problem a brute force enumeration to be preferable to algorithm (K).

EXERCISES

4.1 Solve the inspection problem (example 4.1) given the following data:
$n = 5, B = 2000, b_1 = 0.03, b_2 = 0.02, b_3 = 0.04, b_4 = 0.03, b_5 = 0.03$.

f_{ij}/B_{vj} tabulated

i \ j	1	2	3	4	5
0	16/29	34/30	39/33	44/38	46/42
1		30/18	34/20	40/23	42/26
2			28/17	33/19	38/22
3				27/13	34/16
4					28/11

r_{ij} tabulated

i \ j	1	2	3	4	5
1	0.45	0.90	1.50	2.15	2.60
2		0.50	1.10	1.70	2.10
3			0.70	1.40	1.80
4				0.80	1.35
5					0.60

4.2 (Line-balancing—Klein, 1963) A series production line consists of manufacturing operations numbered $1, 2, \ldots, n$ with operation i requiring a time t_i to perform. Operations are to be formed into s groups $J_\alpha = \{k_{\alpha-1}+1, \ldots, k_\alpha\}$ where $0 = k_0 < k_1 < \ldots < k_s = n$. All operations on an item at station i must be completed within a cycle time C. It is required to determine s and the partition $(J_1, J_2, \ldots, J_s)$ so that the idle time $\sum_\alpha (C - T_\alpha)$ is minimised where $T_\alpha = \sum_{i \in J_\alpha} t_i$. Model this as a shortest route problem through a network which for $s = 5$ has the form of the network of figure 4.1.

4.3 Use algorithm (D) to find the SD-tree shown in figure 1.4.

4.4 A communication network consists of n stations and certain links between stations. Each link ij has (independent) probability p_{ij} of being in a working state, and so the probability of a 'line' $1 = i_1 i_2 i_3 \ldots i_{s-1} i_s = n$ being 'open' between stations 1 and n is

$$P_{i_1 i_2} \cdot P_{i_2 i_3} \cdot \ldots \cdot P_{i_{s-1} i_s}$$

Verify that a line from 1 to n with the highest probability of being open can be found using dynamic programming. Show also that the problem can be treated as a shortest route problem by letting the length of ij be $-\log p_{ij}$. Solve exercise 2.10 as a shortest route problem.

4.5 Show that the knapsack problem

$$\text{maximise} \sum_{i=1}^{n} p_i x_i$$

subject to

$$\sum_{i=1}^{n} t_i x_i \leqslant T \qquad x_i \text{ integer} \qquad (t_i \neq t_j, i \neq j)$$

can be solved by finding the shortest path from vertex 1 to vertex $T+1$ in the network $N = (V, A, w)$ where

$V = \{1, 2, \ldots, T, T+1\}$
$A = A_1 \cup A_2$
$A_1 = \{ab \mid b - a = t_i\} \qquad w(ab) = p_i$
$A_2 = \{a(T+1) \mid \text{all } a \leqslant T\} \qquad w[a(T+1)] = 0$

Compare this with the approach of example 2.12.

4.6 It is required that a circuit through vertex b with the least number of arcs be found in a graph $G = (V, A)$. Define

$$d_{ij} = \begin{cases} 1 & \text{if } ij \in A \\ \infty & \text{otherwise} \end{cases}$$

Algorithm (D) is used to build up an SD-tree (based on vertex b) as usual except that after all nodes distance 1 from the root node b have been added to the tree, vertex b is again allowed as a candidate for addition.

Apply this approach to the housing problem of example 2.6 to find all shortest circuits through vertex 1, then after reallocation of houses and the associated modifications to the desired changes matrix, again find all shortest circuits through vertex 1. Repeat this procedure until no circuits through vertex 1 remain. Then perform the same process for vertex 2, then vertex 3, etc.

Note that the approach to the housing problem just outlined is essentially that described by Wright (1975). He claims it to have the advantage of tending to generate fewer long circuits because, for the earlier vertices there is more freedom of choice and a later vertex j cannot lead to a circuit which includes i if $i < j$.

4.7 It is required to find a circuit through vertex b, containing at least one other vertex, in network $N = (V, A, w)$. The modified SD-tree method of exercise 4.6 is altered by growing two trees rooted at b, one for graph N and one for its reverse N^{-1}. A node, not equal to b, which is on both trees corresponds to a circuit through b. Discuss whether this bi-directional search might have advantages over the uni-directional one described in exercise 4.6.

4.8 The shortest distance matrix and matrix Θ have been found (by Floyd's algorithm say). The length of arc ab is changed. How can D and Θ be correctly updated without repeating the full calculation? (See Dionne and Florian, 1979, for example.)

4.9 How would you modify Floyd's algorithm to operate with networks kept in a linked list? Is this likely to be more or less efficient than the matrix-based algorithm (Fl) if applied to large sparse networks? (See section 10.3.)

4.10 Use algorithms (D) and (Fd) (or Fd$'$) to find a shortest path and shortest distance, between vertices 1 and 7 in the network of figure 4.7. Also find the shortest distance between every pair of points using algorithm (Fl).

5 Location Problems

Situations in which a physical object, or objects, have to be located are of very frequent occurrence. The objects in question will be called *facilities*. A variety of factors have to be taken into account in location problems but a very commonly occurring one is that of 'accessibility', and this will be the feature of special interest in this chapter.

Location problems can be categorised in several ways. The sites which facilities may occupy will be limited to some set F which may be *discrete*, with the possible sites represented by the vertices of a network, or *continuous*. This latter class may be divided into those problems where F is 1-dimensional, and can be represented by the vertices and edges (arcs) of a network, and those, called *problems in the plane* where F is 2-dimensional. We shall be concerned almost exclusively with problems in networks with the major emphasis being on the discrete case.

The number of facilities to be sited may be one (section 5.1) or several (sections 5.2 and 3).

A distinction is often made between *private sector* and *public sector* facilities location problems which largely correspond to the cases of private and public ownership. Private sector problems are concerned with minimisation of cost (or maximisation of profit) and the basic decision is based on a trade-off between cost of establishing and operating facilities, and transportation costs that will be incurred. High transportation costs and low facilities costs favour decentralisation whereas the reverse favours a few large centrally placed facilities. The simplest problem in the discrete case is the simple location problem discussed in section 5.2.

Public sector decisions involve the features of those in the private sector with the additional one that 'goals, objectives and constraints are no longer easily quantifiable' (ReVelle *et al.*, 1970). Surrogate or substitute measures of utility are (usually) used, the primary aim being to gain insight about the system rather than to define solutions directly. Two common surrogates used are total distance travelled and maximum distance travelled, problems to which these are appropriate being called *ordinary* (sections 5.1 and 2) and *emergency* (sections 5.1 and 3) services facilities problems respectively. Examples of ordinary services facilities include post offices, schools, substations in electric power networks and solid waste disposal facilities. Examples of emergency services facilities include fire stations, police stations and hospitals. Of course real life defies strict classification and actual problems will involve some combination of the two aspects of average travel distance and maximum travel distance. This is accounted for by generalising the objective function as in the *centdian* model (section 5.1) and by adding appropriate constraints as in the *maximal covering location problem*. A fuller discussion of private and public sector services location problems is given in ReVelle et al. (1970).

Sometimes variations in travel time (due to rush hours, etc.) may be of importance. This aspect will not be pursued here and interested readers are referred to Handler and Mirchandani (1979).

5.1 SINGLE FACILITY PROBLEMS

In discussing the choice between plane and network models Krarup and Pruzan (1977) state that 'The decisive questions in such cases are often: (1) is the transportation network so well developed in the region under consideration, that a planar formulation is reasonable? (2) Is there a small set of identifiable, possible locations so that a network formulation is reasonable? (3) Are the optimal solutions to a planar formulation readily transferable to a set of possible locations without resulting in a serious increase in the value of the objective function? (4) Are there considerable computational simplifications obtainable in a network or a planar formulation?' A further discussion of this and other methodological decisions is given in Rand (1976).

We first look briefly at some model problems in the plane before passing to the network models.

The *1-median in the plane* problem (1-MPP) (also called the generalised Weber problem) is

1-MPP: minimise $\tau(x, y) = \sum_j w_j D_j(x, y)$
          x,y

where w_j is the weight attached to the jth customer (population, goods required, etc.) situated at (x_j, y_j), and $D_j(x, y)$ is the Euclidean distance from (x_j, y_j) to (x, y). Note that this is just the Steiner problem of section 2.3 if all weights are equal. If $(\tilde{x}, \tilde{y})$ is an optimal solution then

$$\frac{\partial \tau(\tilde{x}, \tilde{y})}{\partial x} = \frac{\partial \tau(\tilde{x}, \tilde{y})}{\partial y} = 0 \tag{5.1}$$

1-MPP may be solved by an iterative procedure in which an initial solution $(x^{(1)}, y^{(1)})$ is selected, the distance $D_j(x^{(1)}, y^{(1)})$ computed and equation 5.1 solved for $\tilde{x}, \tilde{y}$ giving a second approximation $(x^{(2)}, y^{(2)})$. $D_j(x^{(2)}, y^{(2)})$ are now computed and so on. Care must be exercised in the unlikely event of $(x^{(p)}, y^{(p)})$ coinciding with one of the points (x_j, y_j) in which case $\tau(x, y)$ is not differentiable (see Francis and White, 1974). This procedure will converge arbitrarily close to the global optimum but not as fast as the Newton method providing the latter does not oscillate (Krarup and Pruzan, 1977).

The function $\tau(x,y)$ is convex if the distance function is generalised to

$$D_j(x,y) = K\{|x-x_j|^m + |y-y_j|^m\}^{1/m} \qquad K > 0$$

provided $m \geq 1$, and so there is only one local minimum which must thus be a global minimum. The above iterative procedure can thus be used; however, if $m = 1$ the situation simplifies by separating τ into functions of x and y only

$$\tau(x,y) = \tau_1(x) + \tau_2(y)$$

and the solution is found by solving two 1-variable minimisation problems.

The corresponding minimax problem, the *1-centre problem in the plane* (1-CPP), is

1-CPP: $\quad$ minimise $\quad \sigma(x,y) = \max_j [w_j D_j(x,y)]$
$\quad\quad\quad\;\; x,y$

If all w_j are equal, and $D_j(x,y)$ is the Euclidean distance ($m = 2$), then it is clear that (x,y) is the centre of the smallest circle that contains all the points (x_j,y_j) and it can be found readily by the 'geometric' algorithm of Elzinga and Hearn (1972).

1-median Problem

We now turn to the corresponding problems in networks. The *1-median problem* (1-MP) is

1-MP: $\quad$ minimise $\quad \tau(z) = \sum_i w_i d(z,i)$
$\quad\quad\quad\;\;\; z$

where $d(z,i)$ is the shortest distance in the network from z to vertex i and z is constrained to lie at a vertex or on an edge. If z is a vertex then $d(z,i)$ is well-defined (chapter 4). Otherwise suppose z is a proportional distance θ along edge ab from a, then

$$d(z,i) = \min\,[\theta d_{ba} + d(a,i), (1-\theta)d_{ab} + d(b,i)]$$

(If ab is an arc of a directed network then the first term of the minimisation is dropped.)

Although the formulation permits z to be an interior point of an edge the following theorem shows that, for this problem, attention may be restricted to vertices.

Theorem 5.1 (Hakimi, 1964)
1-MP always possesses an optimal solution z at a vertex.

Proof (For undirected networks) Suppose z^* is a point on edge ab. Let A be the set of vertices i such that a shortest chain from z^* to i passes through a, and let B be the set of remaining vertices. Then

$$\tau(z^*) = \sum_{i \in A} w_i d(z^*, i) + \sum_{i \in B} w_i d(z^*, i)$$

$$= \sum_{i \in A} w_i [d(a,i) + \theta d_{ba}] + \sum_{i \in B} w_i [d(b,i) + (1-\theta)d_{ab}]$$

$$= \theta \{ \sum_{i \in A} w_i [d(a,i) + d_{ab}] + \sum_{i \in B} w_i d(b,i) \}$$

$$+ (1-\theta) \{ \sum_{i \in A} w_i d(a,i) + \sum_{i \in B} w_i [d(b,i) + d_{ab}] \}$$

By using the relations

$$d(a,i) + d_{ab} \geq d(b,i), \qquad d(b,i) + d_{ab} \geq d(a,i)$$

we get

$$\tau(z^*) \geq \theta \sum_i w_i d(b,i) + (1-\theta) \sum_i w_i d(a,i)$$
$$= \theta \tau(b) + (1-\theta) \tau(a)$$

Hence $\tau(a) \leq \tau(z^*)$ or $\tau(b) \leq \tau(z^*)$.
That is, to any z^* there is a vertex, i^* say, such that $\tau(i^*) \leq \tau(z^*)$ and the result follows. □

An optimal solution to any 1-median problem on a network can thus be solved by computing $\tau(i)$ for each vertex i and finding a vertex which yields the minimum value. This is probably the best method in general, but for networks in the form of trees there is a simpler method which relies on the following result in which $w(C)$ denotes $\sum_{i \in C} w_i$ for a subset of vertices C.

Theorem 5.2
Let ab be any edge of a tree and A (B) be the set of vertices reachable from a (b) without passing through b (a). Then

$$w(A) \geq w(B) \text{ implies } \tau(a) \leq \tau(b)$$

Proof

$$\tau(a) = \sum_{i \in A} w_i d(a,i) + \sum_{i \in B} w_i [d(b,i) + d_{ab}]$$

$$\tau(b) = \sum_{i \in A} w_i [d(a,i) + d_{ba}] + \sum_{i \in B} w_i d(b,i)$$

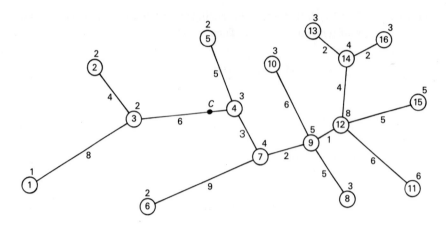

Figure 5.1 Weight w_i is given beside each vertex i, and the length d_{ij} is given alongside each edge ij. C marks the centre

Since $d_{ab} = d_{ba}$

$$\tau(b) - \tau(a) = d_{ab} \left\{ \sum_{i \in A} w_i - \sum_{i \in B} w_i \right\} \geq 0 \quad \square$$

Corollary 1 If $w(A) > w(B)$ and $w_i \geq 0$ all i then B cannot contain an optimal solution to 1-MP. If $w(A) = w(B)$ then a and b are both optimal solutions.

Corollary 2 If $w_i > 0$ for all vertices i of a tree T then T can have at most two vertices which are optimal solutions to 1-MP.

We are now in a position to put forward an algorithm (Goldman, 1969) for finding a 1-median of a tree.

Algorithm (G)
(To find a 1-median of a tree)

Step 1 (Setup)
Form $W = \sum_i w_i$.
Select any vertex j.

Step 2 (Termination)
If $w_j \geq W/2$ then terminate; j is a 1-median.

Step 3 (Iteration)
  If j is a tip vertex (degree = 1)
  then set $w_k \leftarrow w_k + w_j$ where k is the neighbour of j, and discard j;
  else find an elementary chain from j to a tip vertex k using previously unused edges if possible.
  Set $j \leftarrow k$ and return to step 2.

Example 5.1 Find a 1-median for the network of figure 5.1.

Solution $W = 56$ and so $W/2 = 28$.

Step
1 Select vertex 8 as starting vertex, that is $j = 8$.
2 $w_j = 3 < 28$.
3 8 is a leaf. Set $w_9 \leftarrow 5 + 3 = 8$, $j \leftarrow 9$ and discard 8.
2 $w_j = 8 < 28$.
3 9 is not a leaf. Chain 9–12–11 is found. Set $j \leftarrow 11$.
2 $w_j = 6 < 28$.
3 11 is a leaf. Set $w_{12} \leftarrow 8 + 6 = 14$, $j \leftarrow 12$ and discard 11.
2 $w_j = 14 < 28$.
3 12 is not a leaf. Chain 12–15 is found. Set $j \leftarrow 15$.
2 $w_j = 5 < 28$.
3 15 is a leaf. Set $w_{12} \leftarrow 14 + 5 = 19$, $j \leftarrow 12$ and discard 15.
2 $w_j = 19 < 28$.
3 12 is not a leaf. Chain 12–14–13 is found. Set $j \leftarrow 13$.
2 $w_j = 3 < 28$.
3 13 is a leaf. Set $w_{14} \leftarrow 4 + 3 = 7$, $j \leftarrow 14$ and discard 13.
2 $w_j = 7 < 28$.
3 14 is not a leaf. Chain 14–16 is found. Set $j \leftarrow 16$.
2 $w_j = 3 < 28$.
3 16 is a leaf. Set $w_{14} \leftarrow 7 + 3 = 10$, $j \leftarrow 14$ and discard 16.
2 $w_j = 10 < 28$.
3 14 is *now* a leaf. Set $w_{12} \leftarrow 19 + 10 = 29$, $j \leftarrow 12$ and discard 14.
2 $w_j = 29 > 28$ and so the algorithm terminates with vertex 12 as solution.
  (This solution is in fact unique since the inequality is strict.)

The condition that previously unused edges be used in building up the elementary chains in step 3 is to make the search depth-first thus avoiding unnecessary 'jumping about' in the tree. In the above solution vertex 12 was first encountered from edge 9–12 and so 9–12 could not be used again until chains to 11, 15 and via 14 had been formed. However, since 12 is a median it proves to be unnecessary to use 9–12 a second time. Again, vertex 14 is first encountered from edge 12–14 and so 12–14 is not used further until after 14–13 and 14–16 have been used.

1-Centre Problem

If an emergency service facility is to be sited on a network then the 1-centre problem (1-CP) is appropriate.

1-CP: $\quad$ minimise$_z \quad \sigma(z) = \max_i d(z, i)$

Theorem 5.3 (Handler, 1973)
A 1-centre of a network must lie at the mid-point of a longest chain. If the network forms a tree then there is a unique 1-centre.

It is not true in general that there is a vertex which is 1-centre; that is, the analogue of the Hakimi theorem (5.1) does not hold. However, there is an analogue of theorem 5.2 (exercise 5.3) which enables 1-centres of trees to be found in a similar way to that for 1-medians.

We now note that both $\tau(z)$ and $\sigma(z)$ will very often be of interest and the decision may well require $\tau(z)$ to be minimised subject to $\sigma(z) \leq \sigma_0$. An alternative (Halpern, 1976) is to consider a convex combination of $\tau(z)$ and $\sigma(z)$ as objective. This gives the *1-centdian problem* (1-CDP)

1-CDP: $\quad$ minimise$_z \quad \rho(z) = \theta\tau(z) + (1-\theta)\sigma(z) \qquad 0 \leq \theta \leq 1$

and a solution is a *1-centdian*. For the special case in which the network forms a tree the following interesting result holds.

Theorem 5.4 (Halpern)
For a given θ a 1-centdian is located at the 1-centre or at a vertex on the unique elementary chain between the 1-centre and the (nearest) 1-median.

Proof $\quad$ For the proof see Halpern (1976).

Note that, by creating a vertex at the 1-centre if it is interior to an edge, this is a direct analogue of Hakimi's theorem (5.1).

Example 5.2 $\quad$ Find the 1-centdian of the tree of figure 5.1 for all $0 \leq \theta \leq 1$.

Solution $\quad$ It was shown in example 5.1 that the unique 1-median is vertex 12. By inspection, or by the methods of section 4.3, a longest chain is 1–3–4–7–9–12–14–13 with length 26. (1–3–4–7–9–12–14–16 and 1–3–4–7–9–12–11 are alternative longest chains.) The mid-point (of all three chains) is the point C which is 5/6 of the way along edge 3–4 from 3. It is easily checked that

$\tau(C) = 511 \qquad \sigma(C) = 13$
$\tau(4) = 465 \qquad \sigma(4) = 14$
$\tau(7) = 357 \qquad \sigma(7) = 17$
$\tau(9) = 309 \qquad \sigma(9) = 19$
$\tau(12) = 307 \qquad \sigma(12) = 20$

Location Problems

Then a 1-centdian is at

C if $0 \leqslant \theta \leqslant 1/47$
4 if $1/47 \leqslant \theta \leqslant 1/37$
7 if $1/37 \leqslant \theta \leqslant 1/25$
9 if $1/25 \leqslant \theta \leqslant 1/3$
12 if $1/3 \leqslant \theta \leqslant 1$.

For $\theta = 1/47$ every point on the edge C–4 is a 1-centdian; similar results hold at the other break points.

5.2 ORDINARY LOCATION PROBLEMS

As mentioned in the introduction to this chapter the simplest private sector location problem is the simple location problem (SLP), or uncapacitated plant location problem as it is sometimes called.

SLP: minimise $\varphi = \sum_{i,j} c_{ij} x_{ij} + \sum_{i} f_i y_i$

subject to $\sum_{i} x_{ij} = 1 \quad j \in C$ (5.2)

$y_i - x_{ij} \geqslant 0 \quad i \in F, j \in C$ (5.3)

$x_{ij}, y_i \in \{0, 1\}$ (5.4)

where C is the set of customers, F is the set of possible sites for facilities and

$y_i = \begin{cases} 1 \text{ if facility sited at } i \\ 0 \text{ otherwise} \end{cases}$

$x_{ij} = \begin{cases} 1 \text{ if facility at } i \text{ supplies total demand at } j \\ 0 \text{ otherwise} \end{cases}$

c_{ij} is the cost of supplying customer j from i

f_i is the fixed cost of establishing a facility at i

The unit operating cost can vary from site to site but these costs, weighted by customer demands, can be incorporated into the coefficients c_{ij} and need not enter the formulation of SLP explicitly. Also it is assumed that capacity constraints are not binding and so the needs of customer j can be wholly met from a facility i for which c_{ij} is minimal. That is, nothing is lost by assuming x_{ij} to be integer; however, if capacity constraints are present fractional x_{ij} must be allowed for. Rand (1976) argues that first solving an uncapacitated problem will lead to greater insight even if capacity constraints are binding.

SLP can be viewed as a problem on the graph $K_{m,n}$ in which it is required to find a set of n edges, one to each customer, so that the combined weights of these edges ($\sum_{i,j} c_{ij} x_{ij}$) plus the sum of fixed charges for sites with an incident edge ($\sum_{i} f_i y_i$) is minimal. This picture is useful in developing a means of finding good solutions, and bounds in a B & B scheme.

Theorem 5.5
An optimal solution to SLP relative to coefficients (c_{ij}, f_i) is also an optimal solution relative to (c'_{ij}, f_i) where $c'_{ij} = c_{ij} - \lambda_j$.

Proof Any feasible solution has precisely one edge incident to each customer and so its value relative to (c'_{ij}, f_i) is $\Lambda = \sum_j \lambda_j$ less than its value relative to (c_{ij}, f_i). Since the value of each feasible solution is changed by the same amount the result follows. □

Corollary $\sum_j \min_i c_{ij}$ is a lower bound to the optimal solution value.

Proof The result follows by using the transformation $c'_{ij} = c_{ij} - \min_\alpha c_{\alpha j}$ and noting that $c'_{ij} \geq 0$ for all i, j. □

Costs must also be incurred from establishing at least one facility. We will imagine the f_i, instead of being associated with a single site i, are 'shared' between the edges incident at i. Each feasible solution will incur part of the cost f_i, and this can be used to provide a tighter bound.

Theorem 5.6
If $\{\mu_{ij}\}, i \in F, j \in C$ is any set of numbers with

(1) $\mu_{ij} \geq 0$

(2) $\sum_j \mu_{ij} \leq f_i, \quad i \in F$

then $\sum_j \min_i (c_{ij} + \mu_{ij})$ is a lower bound to the optimal value for SLP.

Proof SLP can be expressed as

$$\text{minimise} \quad \varphi = \sum_{i,j} c_{ij} x_{ij} + \sum_i (f_i - \sum_j \mu_{ij}) y_i + \sum_{i,j} \mu_{ij} y_i$$

subject to the constraints of equations 5.2, 3 and 4. Using conditions (1) and (2) and the fact that $y_i \geq x_{ij}$ all j (equation 5.3) we see that

$$\varphi \geq \sum_{i,j} c_{ij} x_{ij} + \sum_{i,j} \mu_{ij} x_{ij} = \sum_{i,j} (c_{ij} + \mu_{ij}) x_{ij} = \varphi_L$$

and so the optimal value, $\min \varphi_L$, to the modified SLP problem with coefficients $(c_{ij} + \mu_{ij}, 0)$ provides a lower bound to the optimal value, $\min \varphi$, of the original

Location Problems 103

SLP. The result now follows from the corollary to theorem 5.5. □

For given λ_j theorem 5.5 corollary is applicable if

$$\mu_{ij} \geqslant \max(0, \lambda_j - c_{ij})$$

and so a strategy for finding a lower bound to φ_L for SLP is to start with $\lambda_j = \min_i c_{ij}$ and increase λ_j for as long as possible subject to $\sum_j \mu_{ij} \leqslant f_i$. An algorithm based on these ideas (Bilde and Krarup, 1977 and Erlenkotter, 1979) now follows. $\min^k{}_j$ will denote the kth smallest element of the set $\{c_{ij}\}$ and e_i will denote $\sum_j \mu_{ij}$ for all $i \in F$.

Algorithm (BKE)
(Lower bound for SLP)

Step 1 (Setup)
Set $k = 1$; $e_i = 0$, $i \in F$; $\lambda_j = \min^k{}_j$, $j \in C$.

Step 2 (Iteration)
For all $j \in C$ perform step 2a.

Step 2a
Set $\epsilon_j = \min\,[\min^{k+1}{}_j - \lambda_j, \min_{\substack{i \\ \lambda_j \geqslant c_i}} (f_i - e_i)]$.

Set $\lambda_j \leftarrow \lambda_j + \epsilon_j$.
Set $e_i \leftarrow e_i + \epsilon_j$ for all i for which $c_{ij} \leqslant \lambda_j$.

Step 3 (Termination)
If $\sum \epsilon_j = 0$ terminate with lower bound equal to $\Lambda = \sum_j \lambda_j$; else set $k \leftarrow k + 1$ and return to step 2.

Example 5.3 A manufacturing company uses a particular raw material at five different sites. It is planned to establish one or more depots which will receive the raw material in bulk and dispatch it to the production sites as required. Four candidate locations for depots are chosen with location i incurring a cost f_i.

Location i	f_i	Delivery cost to site					$\sum_j c_{ij}$
		1	2	3	4	5	
1	7	7	15	10	7	10	49
2	3	10	17	4	11	22	64
3	3	16	7	6	18	14	61
4	6	11	7	6	12	8	44

There is also a transport cost of c_{ij} per ton of raw material shipped from i to a production site j (this including an allowance for the cost of bulk delivery to i, etc.). Where should depot(s) be sited so as to minimise the combined cost (fixed cost plus delivery cost) given the specific data above?

Solution It is readily verified that this problem can be modelled as an SLP. Using algorithm (BKE) gives the following.

Step
1    $k = 1$, $e_1 = e_2 = e_3 = e_4 = 0$.
      $\lambda_1 = 7$, $\lambda_2 = 7$, $\lambda_3 = 4$, $\lambda_4 = 7$, $\lambda_5 = 8$. $\Lambda = 33$.
2a $\epsilon_1 = \min(10-7, f_1 - e_1) = 3$ ($\lambda_i < c_{i1}$ $i \neq 1$).
      $\lambda_1 \leftarrow 10$, $\Lambda \leftarrow 36$, $e_1 \leftarrow 3$ (λ_1 is now greater than c_{11}).
2a $\epsilon_2 = \min(7-7, f_3 - e_3, f_4 - e_4) = 0$ ($\lambda_2 < c_{i2}$ $i \neq 3, 4$).
2a $\epsilon_3 = \min(6-4, f_2 - e_2) = 2$ ($\lambda_3 < c_{i3}$ $i \neq 2$).
      $\lambda_3 \leftarrow 6$, $\Lambda \leftarrow 38$, $e_2 \leftarrow 2$.
2a $\epsilon_4 = \min(11-7, f_1 - e_1) = 4$ ($\lambda_4 < c_{i4}$ $i \neq 1$).
      $\lambda_4 \leftarrow 11$, $\Lambda \leftarrow 42$, $e_1 \leftarrow 7$.
2a $\epsilon_5 = \min(10-8, f_4 - e_4) = 2$ ($\lambda_5 < c_{i5}$ $i \neq 4$).
      $\lambda_5 \leftarrow 10$, $\Lambda \leftarrow 44$, $e_4 \leftarrow 2$.
3    $\sum_j \epsilon_j = 11 > 0$. Do not terminate. Set $k \leftarrow 2$.
2a $\epsilon_1 = \min(11 - 10, f_1 - e_1, f_2 - e_2) = 0$.
2a $\epsilon_2 = \min(15 - 7, f_3 - e_3, f_4 - e_4) = 3$.
      $\lambda_2 \leftarrow 10$, $\Lambda \leftarrow 47$, $e_3 \leftarrow 3$, $e_4 \leftarrow 5$.
2a $\epsilon_3 = \min(6 - 6, f_2 - e_2, f_3 - e_3, f_4 - e_4) = 0$.
2a $\epsilon_4 = \min(12 - 11, f_1 - e_1, f_2 - e_2) = 0$.
2a $\epsilon_5 = \min(14 - 10, f_1 - e_1, f_4 - e_4) = 0$.
3    $\sum_j \epsilon_j = 3 > 0$. Do not terminate. Set $k \leftarrow 3$.
2    $\epsilon_1 = \epsilon_2 = \epsilon_3 = \epsilon_4 = \epsilon_5 = 0$.
3    $\sum_j \epsilon_j = 0$. Terminate. Lower bound $\Lambda = 47$.

The algorithm terminates with $\Lambda = 47$ an increase of 14. With an obvious notation $e = (7, 2, 3, 5)$ and $f - e = (0, 1, 0, 1)$. As all the fixed cost for sites 1 and 3 have been used up in forming the lower bound, we might hope that locating facilities at these two sites would provide a feasible (and hence optimal) solution of value Λ. In fact this solution incurs fixed costs of $7 + 3$ and transportation costs of $7, 7, 6, 7, 10$ for customers $1, \ldots, 5$ giving a total cost of 47. Thus it is optimal to open a facility at site 1 to serve customers 1, 4 and 5 and a facility at site 3 to serve customers 2 and 3.

Not surprisingly algorithm (BKE) often produces a lower bound which is strictly less than the value of an optimal solution; indeed it can happen that there is no set $\{\lambda_1, \lambda_2, \ldots, \lambda_n\}$ for which Λ has the same value as an optimal solution (Bilde and Krarup, 1977 and Geoffrion, 1974). In such cases the solution can be completed by improving the bounds (Erlenkotter, 1979) and using B & B methods if necessary. A natural branching strategy is inclusion/exclusion with which variable y_i, for some i, is fixed at 1 or at 0. Bounds can be found by using algorithm (BKE) with modified data. Thus for exclusion the decision not to use site i ($y_i = 0$) is equivalent to assigning an infinite fixed cost to site i (that is $f_i \leftarrow \infty$). For inclusion the decision to use site i ($y_i = 1$) is equivalent to adding f_i to the initial lower bound and then setting f_i to zero.

Example 5.4 With the data of example 5.3, use algorithm (BKE) to find lower bounds if

(1) it is decided not to use site 1; or
(2) it is decided to use site 2.

Solution (1) f_1 is changed from 7 to ∞. Application of algorithm (BKE) leads to the same changes in λ_j as in example 5.3 until $k = 2$ is reached. Then $\epsilon_1 \leftarrow 1, \lambda_1 \leftarrow 10 + 1 = 11, e_1 \leftarrow 7 + 1 = 8, e_2 \leftarrow 2 + 1 = 3;\quad \lambda_5 \leftarrow 10 + 1 = 11, e_4 \leftarrow 5 + 1 = 6, e_1 \leftarrow 8 + 1 = 9$.

Thus $\lambda = (11, 10, 6, 11, 11)$ is obtained with $\Lambda = 49$, $f - e = (\infty, 0, 0, 0)$ and siting facilities at sites 2 and 4 is easily checked to cost 49. [Note that the cost f_3 is saved by observing that for each j, $\min(c_{2j}, c_{4j}) \leq c_{3j}$.]

(2) The calculation starts with $\Lambda = 3 + \sum_j \lambda_j = 36$ and $f_3 = 0$. It is found, in a similar manner to that described above, that $\lambda = (10, 10, 4, 11, 10)$ with $\Lambda = 48$, $f - e = (0, 0, 0, 1)$. The best solution including site 2 is to establish facilities at the three sites 1, 2, and 3 with total cost 48.

Erlenkotter (1979) gives some computational experience with an algorithm which may be described as algorithm (BKE) with his 'dual adjustment' procedure incorporated for further improving bounds. In particular, an SLP with $n = 100$, $m = 100$ was solved, using an IBM 360/91 computer, for 10 different values of fixed charge $f_i = f$. In each of eight cases the problem was solved within 1 second without branching being required. For the remaining two cases branching was required and just over 3 seconds processing time used.

Suppose now that there is a limitation to not more than p facilities; that is, the constraint $\sum_i y_i \leq p$ is added to the formulation of SLP. This new problem will be denoted by p-SLP.

Theorem 5.7
A lower bound to p-SLP is provided by subtracting $p\lambda_0$ ($\lambda_0 \geq 0$) from (any lower bound to) an *optimal* solution to SLP relative to coefficients $(c_{ij}, f_i + \lambda_0)$.

Proof p-SLP requires the minimisation of

$$\varphi = \sum_{i,j} c_{ij} x_{ij} + \sum_i (f_i + \lambda_0) y_i - \lambda_0 \sum_i y_i$$

$$\geq \sum_{i,j} c_{ij}x_{ij} + \sum_{i}(f_i + \lambda_0)y_p - p\lambda_0 = \phi_L \quad \text{since} \quad \sum_{i} y_i \leq p$$

For any feasible solution, φ is clearly bounded below by the minimal value of φ_L subject to equations 5.2, 3 and 4. The result follows. □

A public sector location problem, closely related to SLP (and p-SLP), is the *p-median problem* (*p*-MP) in which it is required to locate exactly *p* facilities on a network so as to minimise the total of the weighted distances to each customer from a nearest facility

p-MP: minimise $\tau(Z) = \sum_{i} w_i d(Z, i)$
$\quad\quad\quad\quad Z$

$\quad\quad\;\;$ subject to $|Z| = p$

where Z is a set of points on the network and $d(Z, i)$ denotes the distance $\min_{z \in Z} d(z, i)$. It is clear that *p*-MP reduces to the 1-median problem defined in section 5.1 if $p = 1$.

Theorem 5.8 (Hakimi, 1965)
There is an optimal solution Z to *p*-MP which consists of *p* vertices.

Proof Let Z^* be an optimal solution to *p*-MP and suppose that $z^* \in Z^*$ lies on edge ab. Let H be the set of vertices i for which z^* is a nearest vertex of Z^*. Then

$$\tau_H(z^*) = \tau(Z^*) - \sum_{i \notin H} w_i d(Z^*, i) = \sum_{i \in H} w_i d(z^*, i)$$

As in the proof of theorem 5.1, $\tau_H(z^*)$ is equal to $\theta \tau_H(b) + (1-\theta)\tau_H(a)$. Hence

$$\tau(Z^*) \geq \theta \tau[(Z^* - \{z^*\}) \cup \{b\}] + (1-\theta)\tau[(Z^* - \{z^*\}) \cup \{a\}]$$

and thus that there is an optimal solution in which z^* is replaced in Z^* by either vertex a or vertex b.

By repeating this process for each $z^* \in Z^*$, it is seen that there is an optimal solution to *p*-MP consisting solely of vertices. □

Corollary A lower bound to *p*-MP is provided by subtracting $p\lambda_0$ ($\lambda_0 \geq 0$) from (any lower bound to) an optimal solution to SLP relative to coefficients $(c_{ij}, f_i = \lambda_0)$.

Proof By theorem 5.8 the search may be restricted to subsets of vertices containing precisely *p* elements. Now since the fixed cost of a facility does not enter the objective function, *p*-MP is equivalent to *p*-SLP with $f_i = 0$, all *i*. The result now follows from theorem 5.7. □

Example 5.5 A Local Authority plans to establish p new libraries, at sites to be chosen from m potentially suitable ones, in a particular district under its control. The district is notionally split into n smaller units with unit j a distance d_{ij} from potential library site i.

It is believed that the proportion, $\pi(d)$, of persons using a library at a distance d from their homes approximately satisfies (ReVelle and Church, 1977)

$$\pi(d) = \pi_0 \exp(-kd), \qquad \pi_0 \leq 1$$

for appropriate constants π_0 and k. The Authority wishes to locate the libraries so that the maximum (combined) use is made of them. How might this problem be modelled?

Solution Assume to start with that a 'green fields' situation obtains, there being no existing libraries in the district. It is also assumed that a person will use the nearest library or none at all. Then the number of persons in unit j using a library is $w_j \pi_0 \exp(-k \min_{i \in L} d_{ij})$ where w_j is the population of unit j and L is the set of p library sites under consideration. Thus the total number of persons *not* using libraries is

$$\sum_{j=1}^{n} w_j (1 - \pi_0 \exp(-k \min_{i \in L} d_{ij}))$$

$$= \sum_{j=1}^{n} w_j \min_{i \in L} [1 - \pi_0 \exp(-k d_{ij})]$$

$$= \sum_{j=1}^{n} w_j \min_{i \in L} \delta_{ij} = \sum_{j=1}^{n} w_j \delta(L, j)$$

where the distance function δ is defined by

$$\delta_{ij} = 1 - \pi_0 \exp(-k d_{ij})$$

The problem of optimally locating p libraries becomes

$$\text{minimise} \quad \sum_{j=1}^{n} w_i \delta(L, j)$$

subject to $\quad |L| = p$

which is just a p-MP. The same conclusion is clearly reached for other forms of the function π (providing it is positive and monotonically decreasing with distance).

If q libraries are already present then their locations are added to the set of possible sites and a $(p+q)$-MP formed. This problem together with the restriction (which can be incorporated naturally into a B & B scheme) that the q established libraries are to occupy their present sites provides an appropriate model for the problem at hand.

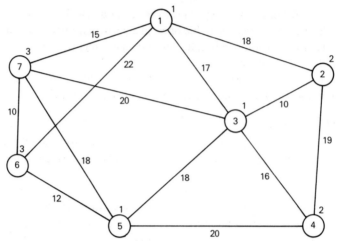

Figure 5.2 The weight w_i is given beside each vertex i, and the length d_{ij} is given alongside each arc ij

Example 5.6 Solve p-MP for $p = 1, 2, \ldots, 6$ given the network of figure 5.2.

Solution The weighted distance matrix $(c_{ij}) = (w_j d_{ij})$ is

$$(c_{ij}) = \begin{bmatrix} 0 & 36 & 17 & 66 & 33 & 66 & 45 \\ 18 & 0 & 10 & 38 & 28 & 120 & 90 \\ 17 & 20 & 0 & 32 & 18 & 90 & 60 \\ 33 & 38 & 16 & 0 & 20 & 96 & 108 \\ 33 & 56 & 18 & 40 & 0 & 36 & 54 \\ 22 & 80 & 30 & 64 & 12 & 0 & 30 \\ 15 & 60 & 20 & 72 & 18 & 30 & 0 \end{bmatrix}$$

Algorithm (BKE) is now applied with $f_i = \lambda_0$ for selected values of λ_0 giving lower bounds $\text{LB}(\lambda_0) = \Lambda - p\lambda_0$ to p-MP. Results are listed in table 5.1.

Maximum values for $\text{LB}(\lambda_0)$ are found at $\lambda_0 = 12, 15, 33, 38$ and 40 for $p = 6, 5, 4, 3$ and 2 respectively. Corresponding values of $(f - e)$ and suggested solution vectors y (obtained by setting $y_i = 1$ if $f_i - e_i = 0$ and $y_i = 0$ otherwise) are shown in table 5.2.

Let $Y = \{i \mid y_i = 1\}$; it is seen that $\tau(Y) = \text{LB}(\lambda_0)$ where the lower bound is calculated for $p = |Y|$ (that is, the number of facilities to which y corresponds). It follows that optimal solutions to this p-MP for $p = 6, \ldots, 2$ are given by locating facilities at the sites in Y.

For $p = 1$ complete enumeration, computing $\tau(i)$ for each i, is the most efficient method. But $\tau(i)$ is just the sum of the elements in row i of (c_{ij}) and it is seen that vertex 7 is the sole 1-median.

Table 5.1

λ_0	$\Sigma\lambda_j$	\multicolumn{5}{c}{$\Sigma\lambda_j - p\lambda_0$}
		$p=6$	$p=5$	$p=4$	$p=3$	$p=2$
0	0	0	0	0	0	0
10	70	10	20	30	40	50
12	82	10	22	34	46	58
15	97	7	22	37	52	67
33	169	−29	4	37	70	103
38	184	−44	−6	32	70	108
40	188	−52	−12	28	68	108

Table 5.2

e		Solution vector y	$\tau(Y)$	$LB(\lambda_0)$
12	(0,0,2,0,0,0,0)	(1,1,0,1,1,1,1)	10	10 for $p=6$
15	(0,0,5,0,3,0,0)	(1,1,0,1,0,1,1)	22	22 for $p=5$
33	(15,0,8,0,18,0,0)	(0,1,0,1,0,1,1)	37	37 for $p=4$
38	(20,0,1,0,20,0,3)	(0,1,0,1,0,1,0)	70	70 for $p=3$
40	(22,0,3,2,22,0,3)	(0,1,0,0,0,1,0)	108	108 for $p=2$

Not surprisingly p-median problems do not always solve as easily as the above example suggests. There may exist a *duality gap* in that for a particular p, $\max_\lambda LB(\lambda) < \min \tau(Y)$, or even if there is no gap y may have more than p nonzero components. Such matters can be resolved, but often at considerable computational expense, by embedding the bounding procedure in a B & B scheme (Erlenkotter, 1979). A value of λ maximising $LB(\lambda)$ can be found by using a Fibonacci search. Using such an approach Boffey and Karkazis (1981) were able to solve a Euclidean p-MP with $n = 206$ and $p = 45$; a time of 72 seconds was required (on an ICL 1906S computer) and the B & B search tree had 50 nodes.

5.3 LOCATION OF EMERGENCY FACILITIES

We look first at the set covering problem which often arises in connection with emergency facility problems.

Example 5.7 (Arabeyre *et al.*, 1969) An airline maintains routes which consist of m flight legs each of which requires a crew. The legs can be grouped into sets $P_1, P_2, \ldots, P_n$ of legs where each set can be handled by a single crew. If crew members are allowed to be passengers when necessary, and c_j is a measure of the cost of set P_j, find a collection $\{P_{i_1}, \ldots, P_{i_r}\}$ of sets which covers all the m legs with minimal total cost.

Solution   Let

$$a_{ij} = \begin{cases} 1 & \text{if leg } i \text{ is in set } P_j \\ 0 & \text{otherwise} \end{cases}$$

$$x_j = \begin{cases} 1 & \text{if set } P_j \text{ is used} \\ 0 & \text{otherwise} \end{cases}$$

Then the problem resolves to the ILP

SCP: minimise $\varphi = \sum_{j=1}^{n} c_j x_j$

subject to

$$\sum_{j=1}^{n} a_{ij} x_j \geq 1 \qquad i = 1, 2, \ldots, m$$

$$x_j \in \{0, 1\} \qquad j = 1, 2, \ldots, n$$

and where $c_j > 0$, $a_{ij} \in \{0, 1\}$.

SCP can often be reduced at the outset by eliminating certain rows and columns (Garfinkel and Nemhauser, 1972). If R_i denotes the ith row $(a_{i1}, a_{i2}, \ldots, a_{in})$ and C_j the jth column $(a_{1j}, a_{2j}, \ldots, a_{mj})^T$ of the matrix (a_{ij}), then

REDUCTION 1:   If R_i is a null vector then SCP is infeasible as item i cannot be covered.

REDUCTION 2:   If R_i is a unit vector whose only non-zero component is the kth (that is, $a_{ik} = 1$) then $x_k = 1$ is in every feasible solution and column k may be deleted. Also every row t with $a_{tk} = 1$ may be deleted.

REDUCTION 3:   If $R_p \geq R_q$ (that is, $a_{pj} \geq a_{qj}$ all j) then row p may be deleted.

REDUCTION 4:   If for some set of columns S and some column k, $\sum_{j \in S} a_{ij} \geq a_{ik}$ for all i, and $\sum_{j \in S} c_j \leq c_k$ then column k may be deleted.

After making such reductions as are available SCP can be solved by a B & B using inclusion/exclusion branching. At node (E, U, I) (see section 3.2) the greatest lower bound is given by the solution to the SCP

minimise $\sum_{j \in U} c_j x_j + \sum_{j \in I} c_j$

subject to

$$\sum_{j \in U} a_{ij} x_j \geq 1 - \sum_{j \in I} a_{ij} \qquad i = 1, 2, \ldots, m$$

$$x_j \in \{0, 1\}$$

A practical lower bound is obtained by relaxing the integrality constraint to $0 \leq x_j \leq 1$ all $j \in U$. This basic scheme can be improved in various ways (Garfinkel and Nemhauser, 1972 and Christofides, 1975).

Computational experience is reported in Christofides and Korman (1975). Recently Balas and Ho (1980) reported experiments with a method which is very effective for sparse matrices. For a set of ten randomly generated problems with $m = 200$, $n = 1000$ and 2 per cent of the elements of matrix A non-zero, times to obtain guaranteed optimal solutions ranged from 13 to 416 seconds with a mean of 111 seconds on a DEC 20/50 computer. For nine randomly generated, 2 per cent dense problems with $m = 200$, $n = 2000$ the corresponding times were 26, 393 and 164 seconds respectively (a tenth problem of this size could not be solved in 600 seconds).

The concept of maximal service time or distance is well established in the regional location of emergency facilities such as fire stations or ambulance dispatching stations (Church and ReVelle, 1974). If s is the maximal service distance then customer j can only be served by a facility at i if $d_{ij} \leq s$. Given s, the problem of locating a minimal number of facilities such that all customers receive service is called the location set covering problem and is clearly the following special case of SCP

LSCP: minimise $\varphi = \sum_j x_j$

subject to

$$\sum_{k \in N(j)} x_k \geq 1 \qquad \text{all } j$$

$$x_j \in \{0, 1\} \qquad \text{all } j$$

where

$$N(j) = \{i \mid d_{ij} \leq s\}$$

is the set of locations from which customer j might possibly be serviced.

It may happen that the facilities that can be provided within the available budget cannot be located so as to enable all customers to be served within a service distance s. In this case the maximum covering location problem (MCLP), which is to locate the facilities so as to satisfy as much of the demand as possible, is relevant. We formulate the problem in terms of minimising unsatisfied demand (Church and ReVelle, 1974)

MCLP: minimise $\sum_j h_j x_j$

subject to

$$\sum_i a_{ij} y_i + x_j \geq 1$$

$$\sum_i y_i \leq p$$

$$x_j, y_i \in \{0, 1\}$$

$$a_{ij} = \begin{cases} 1 \text{ if } i \in N(j) \\ 0 \text{ otherwise} \end{cases}$$

h_j is the weight of customer j

Here y_i is 1 if a facility is sited at i and 0 otherwise, and x_j is 1 if the customer at j is not covered and 0 if he is. By varying p, a tradeoff curve may be developed, and the decision-maker can then make his decision with a knowledge of the extra benefit that may be derived from having one more facility.

Theorem 5.9
Let $\{\lambda_j\}$ be any set of n numbers satisfying $0 \leq \lambda_j \leq h_j$ then

$$\sum_j h_j x_j \geq \sum_j \lambda_j - [\text{sum of } p \text{ largest sums } (\sum_j a_{ij} \lambda_j)] = D(\lambda)$$

Proof

$$\sum_j h_j x_j \geq \sum_j \lambda_j x_j$$

$$\geq \sum_j \lambda_j (1 - \sum_i a_{ij} y_i)$$

$$= \sum_j \lambda_j - \sum_i (\sum_j a_{ij} \lambda_j) y_i$$

$$\geq \sum_j - [\text{sum of } p \text{ largest } (\sum_j a_{ij} \lambda_j)] \quad \square$$

Example 5.8 Solve the MCLP for the network of figure 5.3 given that the budget allows the provision of two facilities capable of covering up to distance 5.

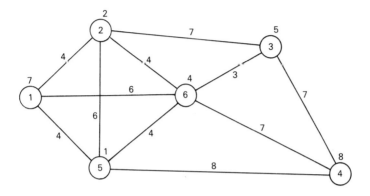

Figure 5.3 The weight w_i is given beside each vertex i, and the length d_{ij} is given alongside each edge ij

Solution First notice that $a_{3j} \leqslant a_{6j}$ for $j = 1, 2, \ldots, 6$ indicating that any site covered by a centre at vertex 3 is also covered by a centre at vertex 6. Thus we need not consider the possibility of a centre at vertex 3 and an equivalent problem is obtained by omitting row 3 of matrix A which becomes

$$A = (a_{ij}) = \begin{bmatrix} 1 & 1 & 0 & 0 & 1 & 0 \\ 1 & 1 & 0 & 0 & 0 & 1 \\ 0 & 0 & 0 & 1 & 0 & 0 \\ 1 & 0 & 0 & 0 & 1 & 1 \\ 0 & 1 & 1 & 0 & 1 & 1 \end{bmatrix} \qquad h = \begin{bmatrix} 7 \\ 2 \\ 5 \\ 8 \\ 1 \\ 4 \end{bmatrix}$$

Try $\lambda = h$. Then

$$A\lambda = \begin{bmatrix} 10 \\ 13 \\ 8 \\ 12 \\ 12 \end{bmatrix} \qquad D(\lambda) = \sum_j \lambda_j - 13 - 12$$
$$= 27 - 25 = 2$$

It may be noticed that reducing λ_6 by 1 reduces $\sum_j \lambda_j$ by 1 and $(A\lambda)_2$, $(A\lambda)_5$ and $(A\lambda)_6$ by 1 each giving a new value of 3 for $D(\lambda)$. Reducing λ_6 by 4 gives

$$\lambda = \begin{bmatrix} 7 \\ 2 \\ 5 \\ 8 \\ 1 \\ 0 \end{bmatrix} \quad A\lambda = \begin{bmatrix} 10 \\ 9 \\ 8 \\ 8 \\ 8 \end{bmatrix} \quad D(\lambda) = 23 - 10 - 9 = 4$$

Reducing λ_1 and λ_2 each by 1 and increasing λ_6 by 1 gives

$$\lambda = \begin{bmatrix} 6 \\ 1 \\ 5 \\ 8 \\ 0 \\ 1 \end{bmatrix} \quad A\lambda = \begin{bmatrix} 7 \\ 8 \\ 8 \\ 7 \\ 7 \end{bmatrix} \quad D(\lambda) = 22 - 8 - 8 = 6$$

There is now no obvious way in which $D(\lambda)$ can be increased and so we take LB = 6.

Now the two largest sums $(\sum_j a_{ij} \lambda_j) = (A\lambda)_i$ are $(A\lambda)_2$ and $(A\lambda)_4$ and (by the concept of complementary slackness — see for example Taha, 1976) we might expect 2 and 4 to be promising sites for locating the two facilities. Accordingly taking $y = (0,1,0,1,0,0)$ it is found that all but communities 3 and 5 are covered leading to a value of $h_3 + h_5 = 6$ for the objective. Since a lower bound of 6 has been obtained it follows that $(0,1,0,1,0,0)$ is an optimal solution.

MCLP does not always solve so simply and if there remains a 'gap' between the values of the best lower bound found and the best solution found then B & B may be used to complete the solution.

p-Centres

Another problem concerning emergency facilities, and one which has attracted considerable attention, is that of finding a *p*-centre. In this problem, which we denote by *p*-CP, it is required to locate *p* facilities on a network (at vertices or on edges) so as to minimise the maximum distance travelled.

p-CP: Find a set C of p points (at vertices or on edges) of a network $N = (X, E, d)$ which minimises

$$\sigma(C) = \max_{i \in X} d(i, C)$$

Any set C which is optimal for p-CP will be termed a *p-centre*. The Hakimi theorem (theorem 5.8) does not extend to p-centres which must, in general, contain interior points of edges. However p-centres can always be found by choosing from a *finite* set B, consisting of the union of the set of vertices X and the set EB of edge-bottleneck points.

Definition 5.1
$\xi = [ij;a,b]$ is an *edge bottleneck point* if ξ is an interior point of edge ij and

(1) ξ is equidistant from the distinct vertices a and b;
(2) the edge segment $i\xi$ between i and ξ (ξj between ξ and j) is contained in no shortest route from ξ to b (from ξ to a).

The distance $d(\xi, a) = d(\xi, b)$ is the *bottleneck distance* of ξ and will be denoted by $d(\xi \mid a, b)$.

It is easily shown that there is at most one edge-bottleneck point $[ij;a,b]$ for given i, j, a and b.

Theorem 5.10
Any network $N = (X, E, d)$ possesses a p-centre C^* with $C^* \subset B = X \cup EB$.

Proof Let C be any p-centre of N. The proof will consist of finding a subset $\tilde{C} \subset B$ with no more than p elements and such that $\sigma(\tilde{C}) \leq \sigma(C)$. $\tilde{C}$ is augmented by arbitrarily adding elements of B until a set $C^* \subset B$ containing p elements is formed. Then $\sigma(C^*) \leq \sigma(\tilde{C}) \leq \sigma(C)$ and the desired result follows.

For each $y \in C$ define

$$S(y) = \{j \mid j \in X, d(j, y) = \min_{x \in C} d(j, x)\}$$

$S(y)$ is just the set of vertices which are as close to y as to any other point in C and can be regarded as the set of vertices that could be assigned to a facility at y. Then if

(1) y is an edge-bottleneck point set $\tilde{y} = y$;
(2) $S(y) = \phi$ then y serves no purpose and no $\tilde{y}$ is formed;
(3) $S(y)$ comprises a single element $\{k\}$ set $\tilde{y} = k$;
(4) $S(y)$ contains several elements then set $\tilde{y}$ to be the 1-centre for the subnetwork on $S(y)$. (By theorem 5.3 this 1-centre must belong to B.)

Set $\tilde{C} = \{\tilde{y} \mid y \in C \,\&\, S(y) \neq \phi\}$. Clearly, for each pair $(y, \tilde{y})$

$$d(j, \tilde{y}) \leq d(j, y) = \min_{x \in C} d(j, x) \quad \text{all } j \in S(y)$$

Consequently $\sigma(\tilde{C}) \leq \sigma(C)$ as required. □

The fact that $y \in C$ is an edge-bottleneck point is relevant if

$$d(y \mid a, b) = d(y, a) = d(y, b) \leq \sigma(C)$$

Thus if $d(y \mid a,b)$ is very large then y is not a candidate as a point of a p-centre ($p > 1$). Correspondingly we define a restricted set

$$B(U) = \{ y \mid y \in B \ \& \ [d(y \mid a,b) \leq U \text{ if } y \notin X]\}$$

$N(U)$ will denote the network obtained from N for which all points in $B(U)$ are regarded as vertices. The following algorithm (Garfinkel et al., 1977) repeatedly takes trial values, b, and tests to see if there is a feasible solution C with $\sigma(C) \leq b$. By varying b, systematically feasible solutions with monotonically decreasing values of $\sigma(C)$ can be obtained.

Algorithm (GNR)
(To approximate a p-centre)

Step 1 (Setup)
    Set $k = 1$, $L = 0$, δ = some specified small positive number and
    $U = \sigma(\tilde{C})$ for any feasible solution $\tilde{C}$ of p-CP.
    Determine the set $B(U)$.

Step 2 (Iteration)
    Set $b = (U + L)/2$.
    If LSCP on network has a feasible solution C with $|C| = p$
    set $\tilde{C} \leftarrow C$, $U \leftarrow \sigma(C)$ if $\sigma(C) < \sigma(\tilde{C})$;
    else set $L \leftarrow b$.

Step 3 (Termination)
    If $U - L \leq \delta$ then terminate with $\tilde{C}$ as approximation to a p-centre;
    else return to step 2.

Example 5.9   Find a 2-centre for the network of figure 5.4.

Table 5.3

Edge-bottleneck point	Name	Bottleneck distance
[12; 1, 2]	6	9
[14; 1, 4]	7	11
[15; 1, 5]	8	7
[15; 1, 4]	9	12
[23; 2, 3]	10	9
[25; 2, 5]	11	10
[34; 3, 4]	12	6
[34; 3, 5]	13	11
[35; 3, 5]	14	9
[45; 4, 5]	15	5

Solution In order to provide a guide a 'rough approximation' to a *p*-centre is found, for example the set C comprising vertex 5 and the mid-point of edge 23. Since $\sigma(C) = 14$ it is unnecessary to generate any edge bottleneck points whose bottleneck distance exceeds 14. This leaves the 10 edge bottleneck points listed in table 5.3 which will be numbered 6 to 15 (see figure 5.4).

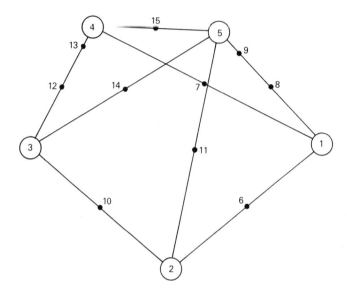

Figure 5.4

Thus $B(14) = \{1, \ldots, 5, 6, \ldots, 15\}$. Now $\tilde{C} = \{5, 10\}$ and $S(5) = \{1, 4, 5\}$, $S(10) = \{2, 3\}$. 10 is certainly a 1-centre as far as $S(10)$ is concerned, but the 1-centre of $S(5)$ is the point 9. $\tilde{C}$ is now updated; $\tilde{C} \leftarrow \{9, 10\}$, and $\sigma(\tilde{C}) = 12$. The iteration starts with $U = 12$, $L = 0$ and $\tilde{C} = \{9, 10\}$.

Step
1 $U = 12$, $L = 0$, $\tilde{C} = \{9, 10\}$, $\delta = \frac{1}{2}$.
2 $b = (12 + 0)/2 = 6$.
   2-CP has no feasible solution with $\sigma(C) \leq 6$.
   $L \leftarrow 6$.
3 $U - L > \delta$. Do not terminate.
2 $b = (12 + 6)/2 = 9$.
   2-CP has no feasible solution with $\sigma(C) \leq 9$.
   $L \leftarrow 9$.
3 $U - L > \delta$. Do not terminate.
2 $b = (12 + 9)/2 = 10\frac{1}{2}$.
   2-CP has no feasible solution with $\sigma(C) \leq 10\frac{1}{2}$.
   $L \leftarrow 10\frac{1}{2}$.

3   $U - L > \delta$. Do not terminate.
2   $h = (12 + 10½) = 11¼$.
    $C = \{6, 13\}$ is a feasible solution to 2-CP, $\sigma(C) = 11$.
    $U \leftarrow 11$, $\tilde{C} \leftarrow \{6, 13\}$.
3   $U - L = ½ = \delta$. Terminate.

Since all edge lengths are integers it follows that the only possible values in the range $(Z, \bar{Z})$ for a 2-centre are 10½ and 11. However, it has been found that 2-CP has no possible solution C with $\sigma(C) \leq 10½$ and so $\{6, 13\}$ must be a 2-centre.

For details of computational experience with this algorithm the reader is referred to Garfinkel *et al.* (1977), but to give an idea of the power of the algorithm we note that these researchers were able to solve a problem with 60 vertices, 174 edges and $m = 12$ in around 2 minutes using an IBM 370/165 computer.

Much of the work involved in the solution of the above problem was related to values of b for which 2-CP had no feasible solution. It may be expected that algorithm (GNR) could be improved by a more adventurous choice for the initial value of L. Finally, we note that algorithm (GNR) tends to be more efficient for values of p for which p/n is relatively large (Garfinkel *et al.*, 1977); an alternative method of Christofides and Viola (1971) exhibits the opposite behaviour in that it tends to be more efficient for values of p for which p/n is relatively small.

EXERCISES

5.1   Prove that if C is a 1-centre of a network N, then C must be the mid-point of a longest chain *through* C. Hence, or otherwise prove the result of theorem 5.3.

5.2   Let a, b, A and B be defined as in theorem 5.2 and let

$$h(A) = \max_{i \in A} d(a, i), \qquad h(B) = \max_{i \in B} d(b, i)$$

Prove that

(i) $h(A) \geq h(B)$ implies $\sigma(a) \leq \sigma(b)$;
(ii) $h(a) > h(b)$ implies no vertex of B and no point on an edge between vertices of B, can be a 1-centre;
(iii) $h(a) = h(b)$ implies the 1-centre is an interior point of edge ab;
(iv) a tree can have only a single 1-centre.

5.3   Use the results of exercise 5.2 to obtain an algorithm, analogous to

Location Problems

algorithm (G), for finding the 1-centre of a tree (Goldman, 1972). Apply your algorithm to the network of figure 5.1.

5.4 Plot the values of $\tau(x)$ and $\sigma(x)$ as x moves from a to e along the linear chain shown in figure 5.5. Observe that τ and σ are convex and piecewise linear, and there is only one point between a and e at which the derivative of σ is not defined. Prove that these results are true for linear chains in general. Do the results remain true if x varies along any chain of a tree?

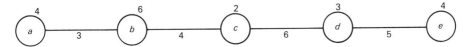

Figure 5.5 The weight w_i is given beside each vertex i and the length d_{ij} is given alongside each edge ij

5.5 A capacitated location problem is obtained from SLP by limiting to s_i the maximum supply from a facility at i, that is, by adding the constraint

$$\sum_j w_j x_{ij} \leq s_i \quad \text{for all } i \in F$$

Show how the fixed costs f_i may be 'shared' between customers by introducing numbers μ_{ij} satisfying the conditions (1) and (2) of theorem 5.6. Hence obtain a lower bound by solving a transportation problem in which the objective $\sum_{i,j}(w_j d_{ij} + \mu_{ij})x_{ij}$ is to be minimised. Illustrate for the problem of example 5.3 if $c_{ij} \equiv w_j d_{ij}$ and

$s_1 = 3 \quad s_2 = 4 \quad s_3 = 3 \quad s_4 = 5$

$w_1 = 1 \quad w_2 = 1 \quad w_3 = 2 \quad w_4 = 1 \quad w_5 = 2$

(For more information on capacitated locations the reader is referred to Nauss (1978) and references contained therein.)

5.6 If r_1 (r_2) is the value of $\sigma(C)$ for C a 1-centre (2-centre) prove that, for a tree

$$\tfrac{1}{4}(2r_1 - 1) \leq r_2 \leq r_1$$

By considering a linear chain and a pair of intersecting chains, or otherwise, prove that both bounds are attained.

5.7 Suppose that a 1-centre of a tree T is an interior point of edge ab. Hakimi et al. (1978) and Handler (1978) have shown that if c_1 and c_2 are the 1-centres of the two trees formed from T by the removal of edge ab then $C = \{c_1, c_2\}$ is a 2-centre for T. Using this result find a 2-centre of the network of figure 5.1.

5.8 (MCLP) The following matrix gives the distances (in miles) between ten cities of the USA together with their populations

	NY	LA	Ch	D	P	SF	B	W	Cl	SL
NY	—	2786	802	637	100	2934	206	233	473	948
LA		—	2054	2311	2706	379	2779	2631	2367	1845
Ch			—	266	738	2142	963	671	335	289
D				—	573	2399	695	506	170	513
P					—	2866	296	133	413	868
SF						—	3095	2799	2467	2089
B							—	429	628	1141
W								—	346	793
Cl									—	529
SL										—
Population ($\times 10^{-5}$)	163	75	71	52	42	37	38	23	23	22

(NY = New York, LA = Los Angeles, Ch = Chicago, D = Detroit, P = Philadelphia, SF = San Francisco, B = Boston, W = Washington, Cl = Cleveland, SL = St Louis.) Solve the corresponding MCLP for values 400 and 600 of the maximum service distance.

6 Project Networks

Projects involving more than a few people, including different specialists, benefit from careful planning and control. This is true whether the project is the construction of a house, the launching of a new product or putting a satellite into space. Any such planning system should (Lock, 1971)

(1) ensure that the project is based on feasible objectives and that individual tasks (or activities) are sequenced in a technologically feasible order;
(2) have a proposed schedule based on *reasonably* reliable estimates for the duration of each task;
(3) be capable of highlighting exceptions: when a task is behind schedule, expenditure is higher or lower than expected, etc.;
(4) be flexible to change when things go wrong;
(5) be matched to available resources, taking into account the requirements of concurrent projects.

For simple projects bar charts may be appropriate, but for more complex ones the critical path method (CPM) is to be preferred. Network aspects of this and related methods are brought together in this chapter, and presented in a graph theory context. For the wider aspects of project management the reader is referred to one of the specialist books (for example, Battersby, 1967, Moder and Phillips, 1970, Lock, 1971 and Burman, 1972). Elmaghraby (1977) may be consulted for a theoretical treatment of the area and for further information on resource allocation.

6.1 CRITICAL PATH METHODS

If a project is to be thoroughly analysed it has to be broken down into component tasks or *activities*. Of course this leaves unanswered the question of what constitutes an activity, that is, to what level of detail should the decomposition be taken. Broadly speaking, it might be expected that the finer the detail considered the more reliable would be the planning. However, it must be borne in mind that project management is concerned with an overall view of the project and that many tactical decisions are best left until the activities in question are about to be performed. For example, in a manufacturing application, the project manager might be concerned with the overall level of loading of a machine shop over various periods of time, whereas the production

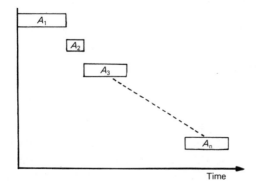

Figure 6.1 A simple sequence of jobs

controller will deal with the sequencing of work on individual machines.

Once the activities have been decided upon, their interrelationships must be specified. For each pair of activities A and B, we need to know whether A must precede B or B must precede A, or A and B can be concurrent. A *simple sequence* of jobs $A_1, \ldots, A_p$ is a sequence such that A_j must precede A_{j+1}, $j = 1, 2, \ldots, p-1$. Such a sequence can be represented by means of a bar chart or as a path $u_0 u_1 \ldots u_p$ where arc $u_{i-1} u_i$ represents activity A_i (figure 6.1). While a bar chart is clearly satisfactory for the consideration of a simple sequence (for example, the workload of a particular team of workers or a particular machine) the network formulation comes into its own in taking account of interconnections of activity sequences. Figure 6.2 gives an example of how three simple sequences can be combined into one network. In this case it was necessary to introduce the *dummy* activity E which is merely used to indicate that A must precede D and will (usually) be assigned a zero time in which to be performed and will receive no allocation of resources.

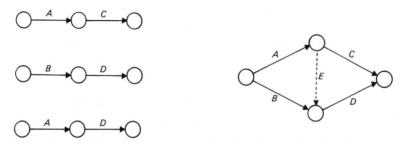

Figure 6.2

Project Networks

This combination of simple sequences can be extended to more complex cases provided suitable dummy activities are incorporated where necessary. A vertex of the composite network, called an *event*, represents the 'milestone' of having completed all activities corresponding to its incoming arcs.

Definition 6.1
A *project network* is a connected directed network without circuits and with a unique source S and a unique sink F. The vertices are called *events*, and the arcs *activities* with the weight $d_{xy} \geq 0$ associated with arc xy being the *estimated duration* of the corresponding activity.

Example 6.1 The building of a bungalow has been split into the activities $A, B, \ldots, U$ and the precedence constraints that are to apply are shown in table 6.1. Construct a project network.

Table 6.1

Name of activity	Description	Must be preceded by	Duration (weeks)
A	Obtain bricks	—	21
B	Obtain concrete	—	3
C	Obtain timber	—	10
D	Obtain electric equipment	—	10
E	Obtain sanitary fittings, etc.	—	17
F	Excavate foundations	—	4
G	Lay foundations	B, F	2
H	Brickwork	A, G	6
J	Lay drains	A, G	2
K	Place roof timbers	C, H	2
L	Complete roofing	K	3
M	Electric wiring	D, L	4
N	Fit exterior doors	K	2
P	Plumbing	E, J, K	4
Q	Plaster	M, N, P	5
R	Joinery	Q	5
S	Fit doors, etc.	R	1
T	Place sanitary fittings	Q	1
U	Point brickwork	S, T	8

Solution An appropriate representation of the above data is given by the network of figure 6.3. Notice that the dummy activity 5–6 arises because the activities preceding 5 (K only) form a subset of the activities preceding 6 (E, J and K).

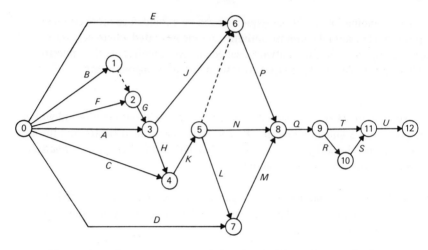

Figure 6.3 The project network for the problem of example 6.1

The condition of connectedness is imposed in definition 6.1 as otherwise each component would be considered a project in its own right. Clearly no circuit can be allowed since this would correspond to a sequence of activities which could not possibly be performed. The requirement of a single source is a convenience rather than a necessity. If there should be more than one source then this can be remedied by selecting one source and inserting a dummy activity from this vertex to each other source (figure 6.4a). Similar remarks apply also to the case of multiple sinks. Finally, it is convenient, if a computer program is to be used, to insist that there is at most one activity between any pair of events. This can be achieved by the incorporation of appropriate dummy activities (figures 6.4b and c).

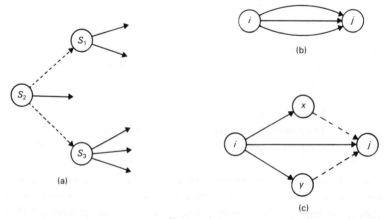

Figure 6.4

Each activity can now be specified uniquely in terms of its end events. For convenience the events will be labelled 0 to n in such a way that $i < j$ for each activity ij. Consequently if there is a path from i to l then $i < l$ and if activity ij must precede activity kl then $j \leq k$. Because of this property, the scheduling of activities can be carried out in an orderly fashion as will be shown later.

A labelling with the above desired property can be found, for any graph without circuits, using the algorithm below.

Algorithm (Fk)
(To label the events of a project network on $n + 1$ vertices)

Step 1 (Setup)
   Set $i = 0$ and $G^{(0)} = G$.

   While $i \leq n$ perform step 2.

Step 2 (Labelling)
   Set $l(\sigma) = i$ for some source σ of $G^{(i)}$.
   Remove σ and all arcs outgoing from σ to form a new graph $G^{(i+1)}$.
   Replace i by $i + 1$.

Step 3 (Termination)
   Each vertex will be denoted by its label.
   Then $i \leq j$ for each activity ij.
   Terminate.

Example 6.2 Use algorithm (Fk) to label the vertices of the project network of example 6.1.

Solution Such a labelling is shown in figure 6.3. In this case the labels are unique except that labels 6 and 7 could be interchanged.

For any vertex x of $G^{(i)}$ either x is a source itself or, since $G^{(i)}$ is finite, there is a path to x which can be traced backwards until a source is found or a circuit formed. But the latter is impossible since G, and hence $G^{(i)}$, does not possess a circuit. Hence it is always possible to select a source in step 2 as required. Also, for any activity xy, $l(x)$ must be set before $l(y)$ and it follows that $l(x) < l(y)$. Thus the algorithm is valid, and from now on we assume it has been applied to all project networks. The algorithm also provides a useful check on the data as the presence of a circuit is detected by the absence of any source for selection in step 2, and only unlabelled vertices can lie on the circuit.

Algorithm (Fk) labels the vertices from 0 to n but any set of distinct numeric labels will suffice as it is the ordering that is important. In practice gaps are often left so that extra activities and events may be incorporated readily should this be required.

Denote by t_i the length, relative to weights d_{xy}, of the longest path from 0 to i then, assuming the project commences at time zero, the following result is obtained.

Theorem 6.1
Activity ij can be scheduled to start as early as (but no earlier than) time t_i, and t_n is the project duration.

Definition 6.2
A longest path from source to sink is a *critical path* of a project network.

The longest path from source to sink (and indeed to any other vertex) is readily found by the following algorithm which, by the principle of optimality, is clearly valid. $ET(t)$ will denote the earliest time at which event i can occur, and A the arc set of the network.

Algorithm (FP)

Step 1
$\quad ET(0) = 0.$

$\quad$ For $j = 1, 2, \ldots, n$ perform step 2.
Step 2
$\quad ET(j) = \max_{ij \in A} [ET(i) + d_{ij}].$

The calculation of the $ET(i)$ is called the *forward pass* and results in a value $ET(n)$ being obtained for the (minimum) *project duration* PD. We are now able to answer the question of *how* late can an event possibly occur without extending the duration beyond PD. This is achieved by the *backward pass* in which $LT(i)$ denotes the latest time at which event i can occur.

Algorithm (BP)

Step 1
$\quad LT(n) = PD.$

$\quad$ For $i = n-1, n-2, \ldots, 0$ perform step 2.
Step 2
$\quad LT(i) = \min_{ij \in A} [LT(j) - d_{ij}].$

This procedure is formally equivalent to making a forward pass in the reverse network and 'working backwards in time'.

From the earliest and latest event time can be derived important constraints on scheduling individual activities

Earliest start time ES(ij) of activity ij — the earliest time at which activity ij can be scheduled to start

$$ES(ij) = ET(i)$$

Earliest finish time EF(ij) of activity ij

$$EF(ij) = ET(i) + d_{ij}$$

Latest finish time LF(ij) of activity ij — this is the latest time at which ij can be scheduled to finish without delaying the completion of the project beyond PD

$$LF(ij) = LT(j)$$

Latest start time LS(ij) of activity ij

$$LS(ij) = LT(j) - d_{ij}$$

Example 6.3 Find the earliest and latest event times for the events of example 6.1, earliest and latest start times for the activities, the project duration, and the critical path.

Solution First set $ET(0) = 0$ then successively calculate

$ET(1) = ET(0) + 3 = 3$
$ET(2) = \max\,[ET(0) + 4, ET(1) + 0] = 4$
$ET(3) = \max\,[ET(0) + 21, ET(2) + 2] = 21$
$ET(4) = \max\,[ET(0) + 10, ET(3) + 6] = 27$
$ET(5) = ET(4) + 2 = 29$
etc.

The full list of earliest event times is given in table 6.2 as also are the lists of latest event times, earliest and latest start times.

It can be seen that there is a flexibility in scheduling the start of activity ij amounting to an interval of time of length $LS(ij) - ES(ij) = LT(j) - ET(i) - d_{ij}$. This quantity is termed the *total float* TF(ij) of activity ij and is the maximum amount of time by which the start of ij could possibly be delayed without delaying the completion of the project. It should be noted however that the start time of ij is dependent on the start times of other activities on the longest path from 0 to n which passes through i and j. Thus for the example of figure 6.5, the total float of activities 6–8 and 8–9 are both equal to 3 time units. However, if 8–9 is scheduled early the start of 6–8 is fixed and conversely if 6–8 is scheduled late. The float of 3 is essentially shared between activities 6–8 and 8–9. A more complicated example is afforded by the float of 1 time unit on activity 2–3.

To get round the 'multiple counting' of the total float of an activity several types of float have been introduced — some are listed below.

Total float TF(ij) of activity ij is the maximum possible time by which the start of ij can be altered

$$TF(ij) = [LT(j) - d_{ij}] - ET(i)$$

Table 6.2

i	ET(i)	LT(i)
0	0	0
1	3	19
2	4	19
3	21	21
4	27	27
5	29	29
6	29	32
7	32	32
8	36	36
9	41	41
10	46	46
11	47	47
12	55	55

Activity	ES(ij)	LS(ij)	TF(ij)	FF(ij)
01	0	16	16	0
02	0	15	15	0
03	0	0	0	0
04	0	17	17	17
06	0	15	15	12
07	0	22	22	22
12	3	19	16	1
23	4	19	15	15
34	21	21	0	0
36	21	30	9	6
45	27	27	0	0
56	29	32	3	0
57	29	29	0	0
58	29	34	5	5
68	29	32	3	3
78	32	32	0	0
89	36	36	0	0
9–10	41	41	0	0
9–11	41	46	5	5
10–11	46	46	0	0
11–12	47	47	0	0

PD = 55

Unique critical path is
0–3–4–5–7–8–9–10–11–12

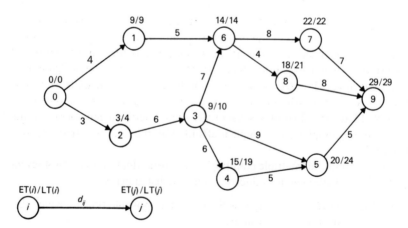

Figure 6.5 TF(35) = 6, FF(35) = 2, SF(35) = 5, IF(35) = 1

Free float   FF(ij) is the maximum possible time by which the start of ij can be altered if subsequent activities are scheduled at their earliest start times

$$FF(ij) = [ET(j) - d_{ij}] - ET(i)$$

Safety float   SF(ij) is the maximum possible time by which the start of ij can be altered if preceding activities are scheduled at their latest start times

$$SF(ij) = [LT(j) - d_{ij}] - LT(i)$$

Independent float   IF(ij) is the maximum possible time by which the start of ij can be altered if preceding activities are scheduled late and succeeding activities scheduled early

$$IF(ij) = \max\,([ET(j) - d_{ij}] - LT(i), 0)$$

Clearly, for any activity ij

$$0 \leqslant IF(ij) \leqslant FF(ij), \quad SF(ij) \leqslant TF(ij)$$

The values of the various floats described are given for activity 35 in figure 6.5.

The construction of a network plays an important part in the analysis of a project and the forward and backward passes provide useful information on scheduling constraints and the amounts of float of the various activities. However, the network is based on estimated durations and assumptions about future conditions. In practice the progress of a project is likely to vary from that planned for various reasons some of which arise externally (for example, weather conditions). If, during the course of a project, progress is out of control (that is, progress differs substantially from that planned), then the original network is readily modified to take account of this. This is achieved by deleting any finished activity and replacing the expected duration of any partly finished activity by an estimate of the time that it will take to finish. If, as is likely, there is now more than one source then dummy activities are added from the source with least label to each of the others (figure 6.4a). It will not matter that the labels do not run from 0 to n as they are still consistent with the precedence constraints.

6.2 ALTERNATIVE APPROACHES

In the previous section activities were represented by arcs. An alternative approach is to represent activities by vertices, and precedence relations by arcs between them.

Example 6.4   A project consisting of four activities U, V, W, T is such that W must be preceded by U and V, and T must be preceded by V. Construct a network model.

Solution Figure 6.6a shows the network with 'activities on arcs'.

Suppose G and $L(G)$ are directed graphs such that the vertex set of $L(G)$ is just the arc set of G and xy is an arc of $L(G)$ if the arcs ij, kl of G corresponding to x and y are such that $j = k$. $L(G)$ is the *arc-to-vertex dual* of G, and for the

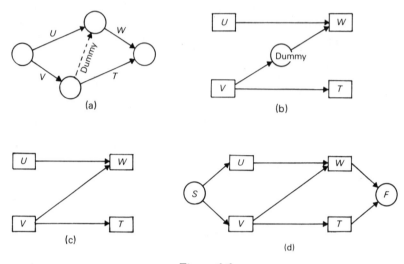

Figure 6.6

graph of figure 6.6a $L(G)$ is shown in figure 6.6b. Clearly the vertex representing the dummy is inessential and can be removed leaving the *activity-on-vertex* network of figure 6.6c, and in figure 6.6d with source and sink vertices added. This highlights one of the advantages of representing activities by vertices, namely, that the troublesome aspect of dummies does not arise. Since a considerable proportion of the activities in activity-on-arc networks can be dummies, representing activities by vertices clearly leads to more compact networks, and construction of networks is simpler. Also the labelling of activities can better reflect the real situation since the ordering aspect is not now necessary; for example, labels could represent a coding in which the first digit of the code defines the responsible department.

If the arcs merely reflect precedence then obtaining earliest and latest start times for scheduling purposes can be carried out in a way analogous to that described in section 6.1. Delays can be represented by assigning weights (times) to the arcs with a negative weight representing overlapping of activities.

Example 6.5 A painting job requires the application of three coats of paint [Primer (*P*), under coat (*UC*) and top coat (*TC*)] to a large surface area which is split into five sections. Each coating of a section takes 2 man days and each section requires 1 clear day between coats. Suppose three men are assigned to the job and that each is responsible for one coat (P, UC or TC) for all the sections. Construct a suitable project network.

Solution This job can be represented by the activity-on-arc network of figure 6.7a, or in the condensed form of figure 6.7b. Condensing networks in this way has to be treated carefully as errors can easily be introduced. Finally an activity-on-vertex network corresponding to the condensed activity-on-arc network is shown in figure 6.7c. Note the negative times on the arcs.

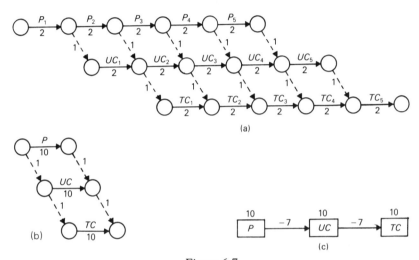

Figure 6.7

Many refinements have been incorporated into the activity-on-vertex representation and interested readers are referred to Burman (1972) for a detailed account. Although the activity-on-vertex networks do have advantages and their use is becoming more widespread, the activity-on-arc representation is still the more common one, and for this reason is adopted for the remainder of this chapter.

Uncertain Activity Durations and PERT

So far it has been implicitly assumed that an activity has a fixed duration which can be estimated accurately, and for some activities these assumptions are reasonable. However, particularly in a research environment, the duration is difficult to estimate accurately when that type of activity has not been performed before. For other activities, whether or not there is a basis of past experience, the duration is intrinsically variable perhaps depending on external influences such as the weather. In such situations it is appropriate to treat the duration as a distributed variable.

PERT (program evaluation and review technique) makes use of three estimates of duration, as perceived by the appropriate specialist, for each activity

d_O = the 'most optimistic' estimate which is such that there is only a 1 in a 100 chance of it being bettered;

d_M = the most likely estimate; that is, the duration which is most likely to occur;

d_P = the 'most pessimistic' estimate which is such that there is only a 1 in a 100 chance of it being exceeded.

From these quantities the expected duration d_E and variance σ^2 of each activity is calculated using the following approximations. (See Battersby, 1967, appendix for derivation and discussion.)

$$d_E = (d_O + 4d_M + d_P)/6$$
$$\sigma = (d_P - d_O)/6$$

The approximation d_E is accurate to within some 10 per cent for a wide range of distributions, but σ is probably less accurate and Moder and Phillips (1970) suggest using $\sigma = (d_{0\cdot9} - d_{0\cdot1})/3$ where $d_{0\cdot9}$ ($d_{0\cdot1}$) is an estimate of the duration which has a 1 in 10 chance of being exceeded (bettered).

PERT finds the estimated project duration PD_E by using the calculations of section 6.1 with activity durations given by d_E. Consider any path π from source to sink, then the expected value of the total time d_π to perform the activities on π is the sum of the d_E for activities on π

$$E(d_\pi) = \sum_{i \in \pi} d_E(i)$$

and the variance of d_π is just the sum of the variances

$$\sigma^2(d_\pi) = \sum_{i \in \pi} \sigma^2(i)$$

provided the durations $d_E(i)$ are statistically independent. If π is the critical path as calculated above then $PD_E = E(d_\pi)$. Also if there are more than a few activities on π then, by a standard statistical result, the distribution for d_π is approximately normal about mean PD_E and with variance $\sigma^2(d_\pi)$. From this and tables of the cumulative normal distribution function the probability that the project can be completed within a certain period, that is $Pr(d_\pi > T)$, can be calculated.

It should be noted that the PERT approach as described involves the implicit assumption that a critical path, calculated by critical path methods using expected durations d_E, is in fact a critical path when the project is undertaken. This is not always true as may be seen from the following example.

Example 6.6 A project consists of five activities A, B, C, D and E whose

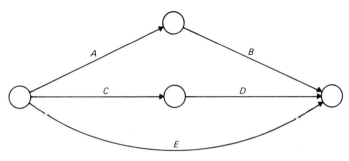

Figure 6.8

logical relationships are shown in the network of figure 6.8. If the activity durations can take only one of a number of discrete values as given in the table what is the critical path?

Activity	A	B			C			D	E
Duration	20	21	24	27	24	26	28	18	45
Probability	1	⅓	⅓	⅓	⅓	⅓	⅓	1	1

Solution Since the distributions for B and C are symmetrical their expected durations can be seen to be 24 and 26 respectively. Using these values the critical path is E with length 45, the expected lengths of paths AB and BC each being 44. However, the length of AB can take values 41,44,47 each with probability 1/3 and CD can take values 42,44,46 again each with probability 1/3. Of the nine combinations of lengths of AB and CD in five of them the maximum of the two *exceeds* 45. That is, if statistical independence between events is assumed, the probability that the critical path is not E exceeds 0.5.

Despite the approximations made and the theoretical deficiencies of PERT it has proved to be quite useful in practice. For a further account of the merits of CPM and PERT and the practicalities involved in their implementation, the reader is referred to the specialist texts.

6.3 RESOURCE ALLOCATION

Time–Cost Tradeoff

The durations of activities were regarded as being fixed for the scheduling calculations of section 6.1. However, they are frequently susceptible to a certain amount of control with allocation of extra manpower, equipment, capital or other resources being made if overall benefit can be gained by doing so.

Example 6.7 It is required that a 6 m long trench be dug and the loose soil removed. Each worker can dig 1 m in 1 h or remove loose soil from 2 m of trench. Discuss what might be an optimal assignment of men, if

(1) congestion effects are taken into account;
(2) each man either digs or carts; and
(3) each man is paid for the duration of the job.

Solution The minimum number of workers is two, in which case the time taken equals the time to dig and the total cost is $12M$ (where M is the cost of 1 man hour). With three men two will dig and one cart, the duration being 3 h for a total cost $9M$. Adding an extra man does not decrease the duration but increases the costs and will be discounted. With five men it is best for three to dig and two to cart giving a duration of 2 h and total cost of $10M$. With six men four will dig and two cart, but now congestion starts to be felt and the digger's average rate of work drops by say $1/7$ m/h. The duration is then $1\frac{3}{4}$ h for a total cost of $10\frac{1}{2}M$. As more men are added congestion becomes worse and there is a limit below which the activity duration cannot be reduced, say $1\frac{1}{3}$ h.

The most economical combination from the point of view of total direct costs is seen to be three men, two to dig and one to cart, taking 3 h. This will be referred to as the *normal* duration, and the minimum duration of $1\frac{1}{3}$ h will be referred to as the *crash* duration.

Since workers come in whole numbers the graph of figure 6.9a consists of a set of discrete points. However, if we allow the men to change between digging and carting in such a way that no one is ever idle then we may assume that the time–cost tradeoff curve is continuous and of the general form shown in figure 6.9b. In practice this curve is frequently replaced by a straight line approximation between the normal and crash durations or, if this is felt to be inadequate, by more than one straight line segment (figures 6.9 c and d).

Sometimes, a continuous line approximation is not adequate. For example, in some trench digging situations there might be the possibility of using a mechanical excavator. In such cases each activity may be replaced by a set of activities of which exactly one must be performed (see exercise 11.8).

The total cost of a project consists of two components

(1) the sum of activity *direct* costs; the direct costs for any activity relate only to that activity and depend on the duration of the activity;
(2) *indirect* costs; these are related to the overall project duration and include overheads such as rents, staff salaries, etc., and penalties through not meeting contracted dates or bonuses for finishing early. This penalty/bonus cost might be explicit payment or, in a maintenance job, loss or gain in time that the equipment is working in relation to scheduled maintenance time.

Reduction of the durations of critical activities leading to a reduction in

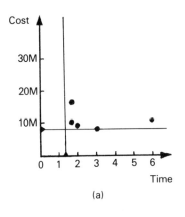
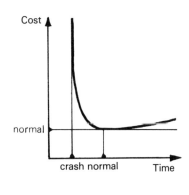
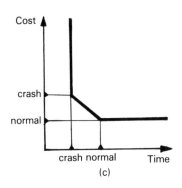
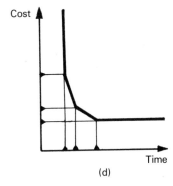

Figure 6.9

project duration is likely to result in increased direct costs but decreased indirect costs. This raises questions such as: 'How should extra resources be allocated so that the total project cost is minimised?' and 'What is the most economical way of meeting a target time T for the project duration?'. The first of these problems can be solved by solving the second for a range of values of T and observing which value leads to minimal total cost. A method of solution will be illustrated by means of an example before being stated formally.

Example 6.8 A project consists of ten activities whose precedence constraints are shown in figure 6.10a. Normal durations, maximum time by which activities can be 'crashed' (that is, their durations reduced) and cost of crashing are given in table 6.3, as also are the calculated total floats corresponding to normal durations. The entries against activity 5–7 indicate that it can be crashed by 1 day at a cost of £10 and by a further day at an extra cost of £20. A saving of

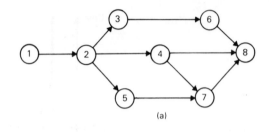

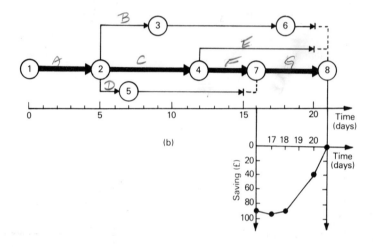

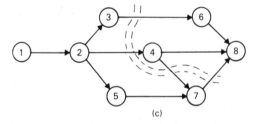

Figure 6.10

£60 in indirect costs can be made for each day by which the project duration is reduced.

What policy of *crashing* (reducing activity durations) should be adopted with a view to minimising the total cost (direct + indirect)?

Solution The network is redrawn to a time scale in figure 6.10b with activities having their normal durations and scheduled as early as possible. The critical path is drawn horizontally and free float is indicated by dashed lines. This arrangement makes it somewhat easier to visualise the changes that are made when crashing.

Table 6.3

Activity	Normal duration (days)	Maximum crashing (days)	Crash cost per day (£)	Initial total float (days)
1–2*	5	2	35	0
2–3	4	1	65	1
2–4*	7	3	30	0
2–5	2	—	—	1
3–6	9	2	15	1
4–7*	4	1	20	0
4–8	8	—	—	1
5–7	8	1	10	1
		1	20	
6–8	2	—	—	1
7–8*	5	1	25	0

It is obvious that the project duration can only be reduced by crashing activities (starred in the table) on the critical path. 4–7 has the lowest crash rate of the critical activities and so it is crashed by 1 day. This increases direct costs by £20 but reduces indirect costs by £60 for a net gain of £40. At this point activities 2–3, 3–6, 6–8 and 2–5, 5–7 and 4–8 became critical; that is, all activities are now critical. From the nature of the network it can be seen that either 1–2 is crashed or 3–6, 2–4 and 5–7 are crashed together. (There is no point in considering 2–3 as its crash rate exceeds £60 per day.) The less costly of these alternatives is to crash 1–2. We crash by its maximum of 2 days for a net saving of 2 × (£60 − £35) = £50. This leaves us with 3–6, 2–4 and 5–7 which we crash by 1 day for a net saving of £60 − (£15 + £30 + £10) = £5. Crashing by a further day would lead to a saving of £60 − (£15 + £30 + £20), that is, a loss of £5 would result.

Thus the best policy found may be summarised as

    Crash 4–7 by 1 day   saving £40
    Crash 1–2 by 2 days   saving £50
    Crash 3–6, 2–4 and 5–7 each by 1 day   saving £ 5

for a total saving of £95, and a new project duration of 17 days.

In order to formalise the above procedure it is convenient to introduce the idea of a 'cut'.

Definition 6.3
A *cut* of a project network is a partition of the set of vertices into two sets A, $\bar{A}$ such that the source belongs to A and the sink to $\bar{A}$. The cut $(A, \bar{A})$ is *crashable* if the

duration of every critical activity ij with $i \in A$ and $j \in \bar{A}$ can be reduced and the duration of every critical activity kl with $k \in \bar{A}$ and $l \in A$ can be extended. The *crash rate*, $cr(A, \bar{A})$, is

$$cr(A, \bar{A}) = \sum_{\substack{i \in A \\ j \in \bar{A}}}^{c} cr(ij) - \sum_{\substack{k \in \bar{A} \\ l \in A}}^{c} cr(kl)$$

where $cr(uv)$ denotes the appropriate crash rate for uv and Σ^c denotes summation over critical activities only. The cut will be termed *positive* if there are no terms in the second summation.

Algorithm (CRASH)

Step 1 (Setup)
  Calculate the total floats of all activities using normal durations.
  Set MINFOUND = *false*.

Perform step 2 for as long as MINFOUND is *false*.

Step 2 (Iteration)
  Select a positive crashable cut $(A, \bar{A})$ with minimal crash rate, if possible else go to step 3. If $cr(A, \bar{A}) > c$ then MINFOUND ← *true*; otherwise crash all critical activities ij with

  $i \in A$ and $j \in \bar{A}$ until

  (1) ij can be crashed no further, or
  (2) the crash rate of ij increases, or
  (3) a non-critical activity becomes critical.

Step 3 (Termination)
  Select that policy, from among the ones considered, with the least total cost.

This is probably sufficiently good for most applications bearing in mind that all activity time–cost tradeoffs have been linearised. That the procedure is heuristic (that is, it does not guarantee an optimal solution) can be seen by considering the cut $(A, \bar{A}) = (\{1,2,3,5,7\}, \{4,6,8\})$ (figure 6.10c) after the crashing described earlier has been carried out. $(A, \bar{A})$ is not positive though it has crash rate

$$cr(A, \bar{A}) = £(15 + 30 + 25) - £20 = £50 < £60$$

Correspondingly 3–6, 2–4 and 7–8 are crashed by 1 day (the maximum possible) and 4–7 uncrashed by 1 day (that is, the duration of 4–7 is put back to its normal duration of 4 days). The project duration has now been reduced to 16 days for a further saving of £10.

The concept of a cut is of fundamental significance in the theory of network flows (chapters 9 and 10), indeed the time–cost tradeoff problem can be formulated as a minimal cost network flow problem and the methods of chapter 10 used for its exact solution (Elmaghraby, 1977).

Other Allocation Problems

Project networks are constructed using only logical restrictions though of course resource availabilities are borne in mind. (For example, for an activity which requires the use of a specialist type of machine the estimate of duration would be based on the use of only one machine if it is known that no more will be available.)

In the above discussion, time was treated as a scarce resource but all other resources such as manpower, machines, materials, and capital were still assumed to be freely available. The following example shows the effect that resource requirements may have on the scheduling of activities.

Example 6.9 Calculate the total daily resource requirements for the network of figure 6.11 if all activities are scheduled to start as early as possible.

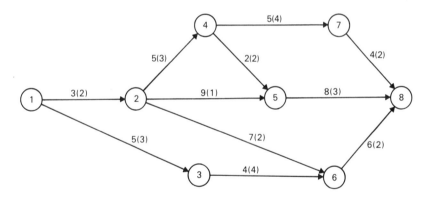

Figure 6.11 The numbers alongside arc ij are of the form $d_{ij}(r_{ij})$ where d_{ij} is the estimated duration and r_{ij} the daily resource requirement

Solution The bar chart of figure 6.12 shows activities (shaded) scheduled at their earliest start times. Total float is shown by the non-shaded part of each rectangle and critical activities are shown cross-hatched. Aggregate resource requirements for each day are shown at the head of the table, and it is seen that the maximum requirement is 13 units on day nine. A project duration of 20 days is attainable if 13 or more units of resource are available.

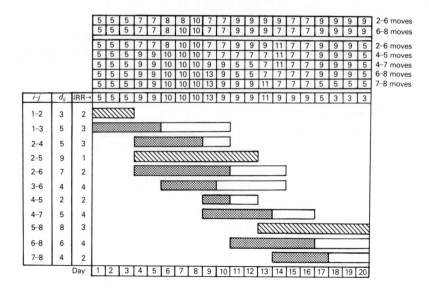

Figure 6.12

Any schedule S will imply a requirement $r_s^\alpha(i)$ for resource α on day i, $i = 1, 2, \ldots, PD_s$, and the vector

$$r_s^\alpha = [r_s^\alpha(1), r_s^\alpha(2), \ldots, r_s^\alpha(PD_s)]$$

will be termed the *profile* of resource α under schedule S. Resource profiles once obtained have to be assessed as to their suitability, which will depend on the circumstances of the particular project. If a schedule S leads to an unacceptable profile for any resource then a new schedule must be sought. Generally, desirable properties of a profile include the following.

(1) The project duration should be as short as possible without incurring excessive costs for resources. On the other hand the maximum resource requirement over any period should not be too high.
(2) The profile should be as smooth as possible without sudden large changes in resource requirement.
(3) $r_s^\alpha(i)$ should approximate $a^\alpha(i)$, the amount of resource α readily available on day i, preferably with $r_s^\alpha(i) \leq a^\alpha(i)$. If $r_s^\alpha(i) > a^\alpha(i)$ then extra resources (overtime, extra personnel or machines, etc.), if available, have to be deployed, usually at a higher rate. A special case arises when $a^\alpha(i)$ is a constant.

Of course an objective function incorporating all these features can easily be devised but the resulting problem is very complex particularly if several resources, crashing and *job-splitting* (that is, breaking off a partially completed activity and finishing it at some later time) are allowed. Thus only heuristic procedures are viable if a range of features is to be considered. Several such procedures have been implemented on computers, for example, SPAR-1 (Wiest, 1967).

Because of the complexity of resource allocation problems attention has been devoted to simple subproblems each of which is appropriate in certain circumstances. Several such subproblems have been identified (for each resource) including the following.

(1) *Resource constrained problem*

$$\text{minimise} \quad PD_s \quad \text{subject to } r_s^\alpha(i) \leq r_0^\alpha$$

(2) *Minimisation of maximum resource requirement*

$$\text{minimise} \quad \max_i r_s^\alpha(i) \quad \text{subject to } PD_s \leq PD_0$$

(3) *Resource levelling*

$$\text{minimise} \quad \sum_{i=1}^{PD_s} r_s^\alpha(i)^2 \quad \text{subject to } PD_s = PD_0$$

(4) *Resource smoothing*

$$\text{minimise} \quad \sum_{i=2}^{PD_s} |r_s^\alpha(i) - r_s^\alpha(i-1)|^c \quad c = 1 \text{ or } 2$$

subject to $PD_s = PD_0$

Even these model problems pose formidable combinatorial difficulties and exact solutions are obtainable only for fairly small problems. A heuristic approach to problems (1)–(3) will now be illustrated by application to the network of example 6.9. Since only one resource is involved the superscript α will be dropped.

Resource Constrained Problem

Example 6.10 Given the network of example 6.9 find the minimum project duration if only 9 units of resource are available on any one day. Crashing and job-splitting are not to be considered at this stage.

Solution A heuristic procedure will be used. At time t the remaining float of an activity ij is defined to be $LS(ij) - t$ where $LS(ij)$ refers to the latest start of ij as computed when resource constraints are ignored.

Algorithm (RA)

Step 1 (Setup)
  Set $t = 0$.

Repeat step 2 while any unscheduled activities remain.

Step 2 (Iteration)

Sort the set, J, of activities whose predecessor activities are completed at time t in order of increasing remaining float (with ties resolved in favour of shorter activities).

Activities of J are considered in order and scheduled to start at time t if the resource level r_0 will not be exceeded.

Set $t \leftarrow$ finish time of the next scheduled activity to finish.

Applying algorithm (RA) to the example network, activities 12, 13, 25, 24, 26 are scheduled at their earliest starts of 0,0,3,3 and 3 respectively (see figure 6.12). 13 will be the next activity to finish and t is set to 5. The only schedulable activity is 3–6, but scheduling it to start at $t = 5$ would result in a resource requirement of 10 units for the next two periods. t is moved on to 8 at which time activity 24 is completed. 45 and 36 have 2 days remaining float and both are scheduled to start. Activity 4–7 must wait to avoid exceeding resource availability. Now t is moved on to 10 at which time 26 and 45 are completed. Activity 47 has only 1 day remaining float compared to 68's 4 days. 47 is scheduled to start at time 10. Activities 2–5 and 3–6 are completed at time 12, and 5–8, which is critical, scheduled to start. t is moved on to 15 at which time 4–7 is completed. 6–8 and 7–8 are now scheduled to start and the procedure terminates. It should be noted that at the last step activity 68 had remaining float of −1 day and consequently finishes a day late at time 21. The minimum project duration, as determined by algorithm (RA) is thus 21 days. The resource profile obtained was

$r = (5,5,5,9,9,6,6,6,9,9,9,9,7,7,7,9,9,9,9,7,4)$

Although the heuristic algorithm (RA) was used it provided an optimal solution in that the target of 20 days is unattainable without crashing, job-splitting or more resources. At this point the project manager might consider crashing one or more activities. Indeed if the critical activity 12 can be crashed by 1 day with daily resource requirement of not more than 6 units then the target of 20 days can be achieved. Alternatively if activity 2–6 can be split then a 20 day project duration can be achieved; the schedule produced by algorithm (RA) is modified by performing the first 2 days of 2–6 as previously scheduled then breaking off at time 5, rescheduling the remainder of 2–6 from $t = 8$, and bringing 3–6, 4–7 and 6–8 forward by 3 days, 1 day and 1 day respectively. The resulting resource profile is

$r = (5,5,5,9,9,8,8,8,9,9,7,7,9,7,7,9,9,9,7,4)$

This simple problem illustrates the general point that the project manager will have more information at his call than can be fed to a computer scheduling program. Consequently, it is desirable to have several good computer produced schedules from which the project manager can choose and on which he can improve by making small adjustments in the light of his extra information.

Project Networks

Methods have been developed for the exact solution of small project duration minimisation problems involving one or more resources.

For more details of the method and for generalisations the reader is referred to Schrage (1970).

Minimisation of Maximum Resource Requirement

Example 6.11   Find, for the network of example 6.9, the minimum resource availability necessary in order to complete the project in 20 days. Crashing and job-splitting are not to be considered at this stage.

Solution   The schedule of figure 6.12 shows that 13 units are sufficient. Also a total of 155 resource unit days are required over 20 days so the average usage per day is $155/20 = 7.75$ units. Since a whole number of units is used at any time the minimum availability is clearly at least 8 units. This lower bound can be improved upon by reconsidering figure 6.11. Since this is the earliest start schedule, better use of the first 3 days cannot be made. Hence at least 140 resource unit days are required over 17 days implying an average daily usage of $140/17 = 8.24$ units. Thus 9 units is a lower bound to the minimum resource availability (see exercise 6.5 for an extension of this idea).

It was in fact shown, in example 6.10, that 9 units is also insufficient. Since a schedule with maximum resource of 10 units in a single day is easily obtainable the minimum resource availability is clearly 10 units.

Generally the strategy is to start from a lower bound on minimum resource availability and keep increasing by 1 unit until a schedule with the required project duration is obtained.

Resource Levelling

Example 6.12   Starting with the schedule of figure 6.12 (p.140) find as level a resource requirement as possible.

Solution   The following heuristic procedure, in which job-splitting is not allowed, will be used.

Algorithm (L)

Step 1 (Setup)
   Number the activities $1, 2, \ldots, m$ so that $i < j$ if activity i must precede activity j.

For $k = m, m-1, \ldots, 2, 1$ perform step 2.

Step 2 (Iteration)
Reschedule activity k so as to make $\varphi = \sum_{i=1}^{PD_s} r_s(i)^2$ as small as possible subject to $r_s(i) \leq r_0$ all i.
In the case of ties use the later schedule.

The starting schedule is shown in figure 6.12 and the value of φ is 1365. The activities will be considered in the order obtained by reading up the diagram of figure 6.12.

Activity 7–8 is rescheduled to start at time 16 and φ is reduced by

$$(9^2 - 7^2) + (9^2 - 7^2) + (9^2 - 7^2) + (3^2 - 5^2) + (3^2 - 5^2) + (3^2 - 5^2) = 48$$

to 1317. 6–8 is rescheduled to start at time 13, φ being reduced by

$$(9^2 - 5^2) + (9^2 - 5^2) + (11^2 - 7^2) + (5^2 - 9^2) + (5^2 - 9^2) + (5^2 - 9^2) = 16$$

to 1301. 5–8 is critical and cannot be moved. Activity 4–7 is rescheduled a day later and φ is reduced by 16 to 1285. Activity 4–5 is rescheduled 2 days later and φ is reduced by 16 to 1269. Moving 3–6 would lead to a higher value of φ. Activity 2–6 is scheduled to start 3 days later and φ is reduced by 4 to 1265. 2–5 is critical and cannot be moved. Moving 2–4 or 1–3 could lead to an increase in φ and 1–2 cannot be moved as it is critical.

Since the activities are considered one at a time it is quite possible that reapplication of algorithm (L) will result in a further reduction in φ. Repeating the process leads to rescheduling activities 6–8 and 2–6 each 1 day later and φ becomes 1245. A further reapplication of algorithm (L) leads to no further improvement. The original and final resource profiles can be found in figure 6.12.

Although this section has only dealt with a single resource the ideas can be extended to the case where several resources are required with an activity possibly using more than one type of resource at once. Allowance for resource requirements of more than a single project can also be dealt with.

For further information on the state of the art of resource allocation problems the reader is referred to Elmaghraby (1977).

EXERCISES

6.1 Discuss how the following projects might be decomposed into activities. What resources might be required and in what quantities? What time factors and constraints might be relevant?

(a) An assault on a previously unclimbed mountain peak is planned to take place in 18 months time when you are appointed to undertake the organisation. You will have to plan the ordering of supplies and equipment,

arrange for shipment of various items, make local arrangements for portering, see that customs and immigration requirements are met, obtain any access permits that are necessary, and attend to many other matters.
(b) The organisation of a political election campaign: you are the party agent and the date of the election has just been announced.
(c) It has recently been discovered that the leaf of a certain plant would provide a convenient substitute for tobacco leaf in cigars. It is believed that this leaf does not lead to the harmful effects attributed to tobacco, but scientific tests to confirm this are still in progress. A new brand of cigar using the substitute leaf is anticipated and an early launch is desirable. You are asked to make plans for such a launch on the premise that the results of the scientific tests are likely to be known in time and that they will be favourable.
(d) You are responsible for the organisation of a major three-day conference. (For the purposes of the exercise you may assume that the topic of the conference is any appropriate one in which you have an interest.)

6.2 Perform forward and backward passes for the network of example 6.1. Check your answers against the results given in table 6.2.

6.3 A project has been decomposed into ten activities $A, B, \ldots, J$, data for which are contained in the table below.

Activity	Must be preceded by	Estimated normal duration (days)	Crashing cost per day (£)	Maximum crashing allowed (days)
A	—	5	70	2
B	J	7	70	3
C	F	6	80	3
D	F	5	20	2
E	H,B,D	3	—	0
F	—	4	90	1
G	A	3	—	0
H	C,G	4	90	2
I	B,D	4	—	0
J	—	4	100	1

Draw an appropriate activity-on-arc network and, using the normal durations, determine for each activity its earliest start, latest start and total float. What is the corresponding project duration? Can the project be completed in this time if each activity requires one work crew, three crews are available, and once an activity has been started it must be carried through to completion without interruption?

If there is a bonus of £250 for each day by which the project duration is reduced what crashing policy would you recommend if the work crew allocation is now ignored? What would the net saving be?

(LU 1977, modified)

6.4 (a) Given the data below for a project network find, for each activity, its earliest and latest start times and total float. Determine the critical path(s).

(b) If a bonus of £100 is payable for each week by which the project duration is reduced what policy for crashing activities would you recommend, and what would the expected saving be?

Activity	Estimated duration (weeks)	Crashing cost (£/week)	Maximum crashing (weeks)
12	5	—	0
13	4	50	1
14	5	—	0
26	5	—	0
34	2	—	0
35	6	65	2
36	9	—	0
45	4	—	0
56	7	70	2
57	3	—	0
68	4	45	2
78	5	60	2

(LU 1978, modified)

6.5 Let E, L and S denote respectively the earliest start, latest start and an arbitrary schedule, all corresponding to a project duration of D. Verify that

$$\sum_{i=1}^{t} r_L(i) \leq \sum_{i=1}^{t} r_s(i) \leq \sum_{i=1}^{t} r_E(i)$$

with all sums equal to R, the total resource requirement, if $t = D$. If u is the first value of i for which $r_E(i) > r_0$ and v is the last value of i for which $r_L(i) > r_0$

prove that no feasible schedule S can exist with duration D and maximum resource availability of r_0 if

$$R > Dr_0 - \sum_{i=1}^{u-1} [r_0 - r_E(i)] - \sum_{i=v+1}^{D} [r_0 - r_L(i)]$$

6.6 Suppose that a schedule S of duration 20 days can be found for the project of example 6.9. Verify the truth of the following assertions.

(a) If there is no job-splitting then activities 24, 25 and 26 must all be in progress on day 8, and so 36 cannot start before day 9;
(b) Activities 47 and 68 may not overlap and so 47 must be in progress on days 10 to 13;
(c) Activities 36, 47 and 58 cannot be in progress together and so 36 must finish not later than day 12;
(d) Statements (a), (b) and (c) together imply that activities 25, 26, 36 and 47 are all in progress on day 12. Consequently the original assumption that schedule S exists must be false.

6.7 Find the critical, and second and third most critical paths for the project network in figure 6.13. Note that the Kth most critical path is the Kth longest path from vertex 0 to vertex 7. (See section 4.3.)

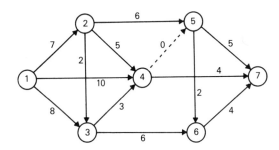

Figure 6.13 The estimated duration d_{ij} is given alongside each activity ij

7 The Travelling Salesman and Chinese Postman Problems

A problem which has attracted much attention, and continues to do so, is the travelling salesman problem (TSP) in which a salesman starting from town T_1 visits $n-1$ other towns $T_2, T_3, \ldots, T_n$, each once only, finally returning to T_1. If the distance d_{ij} from town T_i to T_j is given for each pair (i,j) find a route for the salesman which minimises the total distance travelled. Much of the interest in the problem arises from the challenge presented by the fact that the problem is easy to state and understand but computationally very difficult to solve. Indeed TSP is 'at least as difficult to solve' as any *NP*-complete problem (see appendix).

Although the TSP as stated above is more of a puzzle than a practical problem, the same (abstract) problem does have important applications. It is relevant in collection and delivery problems where a fixed route is maintained, for example, the emptying of mail boxes or carrying children to and from school (Krolak *et al.*, 1971). TSPs with side constraints arise in distribution problems (chapters 1 and 8), and scme scheduling problems can be modelled as TSPs (for an example see exercise 7.8). A variety of approaches to TSP are given in sections 7.1 and 2.

An Euler cycle (circuit) is a cycle (circuit) which, rather than visiting each vertex just once, traverses each edge (arc) of a graph once and once only. If no such cycle (circuit) exists then it is appropriate to ask for one which includes each edge (arc) at least once and which has minimal total length. This is the so-called Chinese postman problem (CPP) (Kwan, 1962) which is relevant to problems relating to refuse collection, street delivery of milk or post, or inspection of electric power, telephone or railway lines (Christofides, 1975), conducted tours, etc. CPP can be solved quite easily by solving a subsidiary matching problem; this is described in section 7.3.

7.1 REDUCTION-BASED METHODS FOR SOLVING TSP

The (asymmetric) travelling salesman problem introduced informally above will now be stated formally.

ATSP: For given *cost matrix* (c_{ij})

   minimise $(c_{1i_1} + c_{2i_2} + \ldots + c_{ni_n})$

where $(i_1, i_2, \ldots, i_n)$ is a permutation of $(1, 2, \ldots, n)$ which is not a product of two smaller permutations.

Let $G = (V, A, c)$ denote the complete directed network on vertex set $\{1, 2, \ldots, n\}$ with the length of (or cost of traversing) arc ij being c_{ij}. ATSP can clearly be restated as follows

ATSP':   Find a minimal length Hamilton circuit of G.

Definition 7.1
A Hamilton circuit (cycle) in a directed (undirected) graph is a circuit (cycle) which visits each vertex once and once only. (Note that, in the context of travelling salesman problems, Hamilton circuits (cycles) are often called *tours*.)

From now on we assume that

(1) $c_{ii} = \infty, i = 1, 2, \ldots, n$
(2) $c_{ij} \geq 0, 1 \leq i \neq j \leq n$.

Clearly (1) imposes no extra constraint since a Hamilton circuit is sought. Also no generality is lost by assuming (2) as can be seen from the following result.

Theorem 7.1
A tour, optimal with respect to cost matrix (c_{ij}), is also optimal with respect to cost matrix (c'_{ij}) where $c'_{ij} = c_{ij} - p_i - q_j$ for all $1 \leq i, j \leq n$, and p_i and q_j are arbitrary real numbers.

Proof   Let $\pi \equiv i_1 i_2 \ldots i_n i_1$ be any tour. Then, since there is just one arc of π into each vertex k and just one arc out of k, the cost of π with respect to (c'_{ij}) is $\sum_k (p_k + q_k)$ less than its cost with respect to (c_{ij}). Because the cost of every tour is reduced by the same amount an optimal tour remains optimal. □

When (c_{ij}) is transformed to (c'_{ij}) as above we say that the cost matrix has been *reduced* by $R = \sum_k (p_k + q_k)$ and that row i and column j have been *reduced* by p_i and q_j respectively. Moreover a transformation (reduction) is said to be *admissible* if p_i and q_j are such that $c'_{ij} \geq 0, 1 \leq i, j \leq n$.

Theorem 7.2
$R = \sum_k (p_k + q_k)$ is a lower bound to the value of an optimal solution to the TSP with respect to (c_{ij}) if the transformation $c'_{ij} = c_{ij} - p_i - q_j$ is admissible.

Proof   Since $c'_{ij} \geq 0$ the cost of an optimal solution of TSP with respect to cost matrix (c'_{ij}) must be non-negative. The result follows. □

Corollary $\sum (\min_j c_{ij}) + \sum_j \min_i (c_{ij} - \min_\alpha c_{i\alpha})$ is a lower bound to the cost of an optimal solution to TSP with respect to (c_{ij}).

If this bound can be extended to subsets of all solutions to a TSP, defined by specifying that certain arcs be included in, and certain arcs excluded from a tour, then a framework for a B & B method has been set up. To do this we first look at a single arc ij for which $c_{ij} = 0$. Let LB, LB($\overline{ij}$) and LB(ij) be bounds for, respectively, the 'parent' TSP, the subproblem with ij excluded and the subproblem with ij included.

(a) *Arc ij is excluded*

Since ij does not belong to any tours the value of c_{ij} cannot affect the cost of any tour. Consequently we set $c_{ij} = \infty$ giving the possibility of reducing row i by $\alpha(i) = \min_{k \neq i} c_{ik}$ and column j by $\beta(j) = \min_{k \neq j} c_{kj}$. Set

$$\text{LB}(\overline{ij}) = \text{LB} + \alpha(i) + \beta(j)$$

(b) *Arc ij is included*

This time there can be no arc iq for $q \neq j$, no arc pj for $p \neq i$ and, if $n > 2$, no arc ji since a Hamilton circuit is sought. Thus we may set each of the corresponding elements of (c_{ij}) to ∞ giving

$$\begin{array}{c} \\ i \\ \\ j \\ \end{array} \begin{array}{c} i j \\ \left[\begin{array}{ccccccc} & & \infty & & & & \\ & & \vdots & & & & \\ & & \infty & & & & \\ \infty \ldots \infty \ldots \infty & 0 & \infty \ldots \infty & \\ & & \infty & & & & \\ & & \vdots & & & & \\ & & \infty & \infty & & & \\ & & \vdots & & & & \\ & & \infty & & & & \end{array} \right] \end{array}$$

Set LB(ij) = LB + R where R is the amount by which this matrix is reduced by an admissible transformation. Note however, that no (positive) contribution to R can be obtained by modifying row i or column j and so they may be deleted leaving an $(n-1, n-1)$ cost matrix. If row j and column i are thought of as describing the same entity then a TSP of smaller size has been obtained. This elimination of row i and column j and setting $c_{ji} = \infty$ corresponds to 'shrinking' arc ij and coalescing vertices i and j in the network in question. Correspondingly row j and column i will be renamed row $\langle i,j \rangle$ and column $\langle i,j \rangle$ respectively.

Since in both cases (a) and (b) a new TSP arises, the above is readily extended to the case where several arcs are sequentially included or excluded. We are now in a position to put forward the method of Little *et al.* (1963) which may be described as a B & B method with

(1) inclusion/exclusion branching;

The Travelling Salesman and Chinese Postman Problems 151

(2) initial bound LB found as in theorem 7.2 corollary, and subsequent bounds calculated as indicated in (a) and (b) above;
(3) for arc ij to be selected for branching on, c_{ij} must be 0 and $\theta(i,j) = \alpha(i) + \beta(j)$ maximal;
(4) a 'change when necessary' development rule (see p.55) is used with preference given to 'inclusion nodes'.

Example 7.1 Solve the TSP with cost matrix

$$(c_{ij}) = \begin{bmatrix} \infty & 10 & 6 & 1 & 9 \\ 13 & \infty & 3 & 12 & 6 \\ 13 & 8 & \infty & 9 & 6 \\ 8 & 16 & 18 & \infty & 12 \\ 25 & 20 & 4 & 13 & \infty \end{bmatrix}$$

Solution Firstly row minima 1,3,6,8 and 4 are subtracted from rows 1,2,3,4 and 5 respectively. The column minima are now 0,2,0,0 and 0 for columns 1 to 5. These are subtracted from the corresponding columns to give a total reduction of 24 (hence initial lower bound LB = 24), and reduced matrix

i \ j	1	2	3	4	5	$\alpha(i)$
1	∞	7	5	0	8	5
2	10	∞	0	9	3	3
3	7	0	∞	3	0	0	(A)
4	0*	6	10	∞	4	4
5	21	14	0	9	∞	9
$\beta(j)$	7	6	0	3	3	

$\alpha(i)$ and $\beta(j)$ are shown for each row and column and it can be seen that $\theta(i,j) = \alpha(i) + \beta(j)$ is maximised for arc 41. Branching is on inclusion/exclusion of arc 41 and $LB(\overline{41}) = 24 + 11 = 35$. To find LB(41) first set $c_{14} = \infty$ delete row 4 and column 1

	2	3	⟨4,1⟩	5
⟨4,1⟩	7	5	∞	8
2	∞	0	9	3
3	0	∞	3	0
5	14	0	9	∞

Row ⟨4,1⟩ reduces by 5 and column ⟨4,1⟩ by 3 to get

	2	3	⟨4,1⟩	5	$\alpha(i)$
⟨4,1⟩	2	0	∞	3	2
2	∞	0	6	3	3
3	0	∞	0	0	0
5	14	0*	6	∞	6
$\beta(j)$	2	0	6	3	

giving LB(41) = 24 + 5 + 3 = 32. Arcs 3⟨4,1⟩ and 53 both lead to maximal $\theta(i,j)$. Select 53 (readers might like to select 3⟨4,1⟩ instead as an exercise). Set $c_{35} = \infty$, delete row 5 and column 3 and reduce row ⟨4,1⟩ by 2 and row 2 by 3 to get

	2	⟨4,1⟩	⟨5,3⟩	$\alpha(i)$
⟨4,1⟩	0	∞	1	1
2	∞	3	0	3
⟨5,3⟩	0	0	∞	0
$\beta(j)$	0	3	1	

It is easily seen that the optimal solution to this TSP on three vertices is ⟨4,1⟩2⟨5,3⟩⟨4,1⟩ (figure 7.1). However, for illustrative purposes we persist with the formal solution one more step and this leads to branching on arc 2⟨5,3⟩. Setting $c_{⟨5,3⟩2} = \infty$ and deleting row 2 and column ⟨5,3⟩ leads to the matrix

	⟨2,5,3⟩	⟨4,1⟩
⟨4,1⟩	0	∞
⟨2,5,3⟩	∞	0

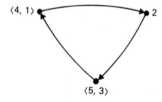

Figure 7.1

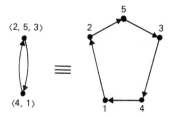

Figure 7.2

and the solution shown in figure 7.2.

The B & B tree grown so far is given by the solid portion of figure 7.3a. It is seen that either the solution 125341 (value 37) is optimal or there exists a better solution with arc 41 excluded. That is, we must backtrack to the node labelled $\overline{41}$. The corresponding TSP has the (reduced) cost matrix (A) with c_{41} set to ∞. This matrix reduces by 11 (as we expect) and arc 53 is selected for branching on. The two nodes generated have bounds of 43 and 38 (figure 7.3a).

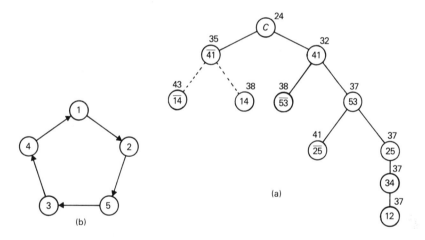

Figure 7.3 (a) The B & B search tree for example 7.1, and (b) the optimal tour

It is thus seen that 125341 is an optimal solution with cost (relative to the original cost matrix)

$$c_{12} + c_{25} + c_{53} + c_{34} + c_{41} = 10+6+4+9+8 = 37$$

in agreement with the value obtained via the reductions.

It should be noted that the shortest arc from each of vertices 1, 2 and 3 was not in the optimal solution.

The bound obtained by the reduction scheme used in the example does not necessarily give a maximum reduction. Thus if the column minima of 8, 8, 3, 1 and 6 were first subtracted from columns 1 to 5 respectively of the original matrix then 1 subtracted from row 5, a reduction of 27 is achieved. As may be expected 'reduce columns before rows' sometimes leads to larger reductions than 'reduce rows before columns' and sometimes to smaller ones. Often, different ways of choosing p_i and q_j can lead to larger reductions still.

How can we find the maximal reduction? Referring back to theorem 7.1, it is seen that the proof relies only on having a permutation π and not at all on the condition that it should not be a proper product of permutations. Geometrically this means that the condition that just one arc of a solution enters and just one arc leaves each vertex is used but it is not required that the solution consist of a *single* circuit. This relaxation of TSP is the (linear) assignment problem (AP) which we now define formally:

AP: For a given cost matrix (c_{ij})

minimise $(c_{1i_1} + c_{2i_2} + \ldots + c_{ni_n})$

where $(i_1, i_2, \ldots, i_n)$ is a permutation of $(1, 2, \ldots, n)$.

A solution corresponds to a set of disjoint circuits in G (defined as for ATSP) such that every vertex is on some circuit. Though generally solutions are allowed to contain loops, in this discussion on TSP we restrict ourselves to the subclass of problems in which loops are excluded by assuming that $c_{ii} = \infty$, $i = 1, 2, \ldots, n$. By the proof of theorem 7.1 it can be seen that we may again assume that $c_{ij} \geqslant 0$, $1 \leqslant i, j \leqslant n$ with no loss of generality.

Theorem 7.3
The cost of an optimal solution to AP is equal to the maximal reduction obtainable using an admissible transformation of the cost matrix (c_{ij}). (For proof see exercise 7.3.)

It has been shown (Dinic and Kronrod, 1969, and Tomizawa, 1971) that AP can be solved in $O(n^3)$ operations, though in practice the dependence on size is often more nearly quadratic (Desler and Hakimi, 1969). Since APs can be solved efficiently (see section 10.3), they could be used to obtain bounds which are somewhat stronger than those used in the method of Little et al. (1963). However, we will describe a somewhat different way of using AP due to Eastman (1958).

Example 7.2 Solve the TSP of example 7.1.

Solution Reducing columns 1 to 5 by 8,8,3,1 and 6 respectively and subtracting 1 from row 5 leads to total reduction of 27 and reduced matrix

The Travelling Salesman and Chinese Postman Problems

$$\begin{bmatrix} \infty & 2 & 3 & 0 & 3 \\ 5 & \infty & 0 & 11 & 0 \\ 5 & 0 & \infty & 8 & 0 \\ 0 & 8 & 15 & \infty & 6 \\ 16 & 11 & 0 & 11 & \infty \end{bmatrix}$$

This cost matrix gives a zero cost solution to AP, namely 141 and 2532, which must be optimal. (See figure 7.4.)

Now consider the circuit 141 (or *subtour* as such components of AP solutions are often called). Both 14 and 41 cannot be in any feasible tour. Consequently the set of all tours can be split into the union of two (not disjoint) sets $X_{\overline{14}}$, $X_{\overline{41}}$ defined by the two cost matrices with c_{14} and c_{41} set to ∞ respectively. After reductions of 10 and 11 respectively, the cost matrices become

$$\begin{bmatrix} \infty & 0 & 1 & \infty & 1 \\ 5 & \infty & 0 & 3 & 0 \\ 5 & 0 & \infty & 0 & 0 \\ 0 & 8 & 15 & \infty & 6 \\ 16 & 11 & 0 & 3 & \infty \end{bmatrix} \quad \begin{bmatrix} \infty & 2 & 3 & 0 & 3 \\ 0 & \infty & 0 & 11 & 0 \\ 0 & 0 & \infty & 8 & 0 \\ \infty & 2 & 9 & \infty & 0 \\ 11 & 11 & 0 & 11 & \infty \end{bmatrix}$$

both of which have zero cost solutions to AP and, since these both consist of but one circuit, are zero cost solutions to TSP. Hence 125341 (145321) is an optimal tour in $X_{\overline{14}}$ ($X_{\overline{41}}$) with cost 37 (38). Since $X_{\overline{14}} \cup X_{\overline{41}}$ is the set of all tours 125341 is an optimal solution to TSP (as found in example 7.1).

The Eastman method is a relaxed form of B & B in which development of a node requires a set of solutions to be split into not necessarily disjoint sets so as to form a covering (see definition 2.18). This can lead to inefficiencies through the search tree being larger than necessary.

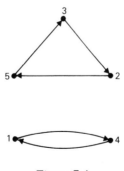

Figure 7.4

This difficulty is readily overcome (in the spirit of Held and Karp, 1970) as follows. Consider any set of arcs $A = \{a_1, \ldots, a_q\}$ such that $a_1, \ldots, a_{q-1}$ can occur simultaneously in some tour, but $a_1, \ldots, a_{q-1}, a_q$ cannot. Then branching takes place to nodes $P_1, \ldots, P_q$ defined by

P_1 : a_1 is excluded
P_2 : a_1 is included a_2 is excluded
P_3 : a_1, a_2 are included a_3 is excluded
         ⋮   ⋮
P_q : $a_1, \ldots, a_{q-1}$ are included a_q is excluded

The principal remaining weakness of the above approaches is that the bounds can be rather weak. This is particularly the case if the vertices form clusters such that a vertex is near to every other vertex in its cluster, but far from every vertex in other clusters. Optimal solutions to the associated AP will comprise a set of circuits with each circuit entirely 'within a cluster'. The AP bound is based entirely on small c_{ij} whereas any tour must contain several larger values of c_{ij}. Christofides (1972) introduced an ingenious way of improving the bounds in such cases, for a relatively small amount of effort. A solution to an AP naturally partitions the set of vertices into clusters $V_1, \ldots, V_r$ with each cluster corresponding to a single circuit. (For example, the solution 141, 2532 to the AP in example 7.2 leads to the clusters $\{1,4\}$, $\{2,5,3\}$.) Now every feasible solution to TSP has at least one arc entering each V_i and at least one arc leaving. In order to obtain a lower bound to the total cost of inter-cluster arcs only the shortest arc from V_i to V_j, all $i \neq j$ is retained, then a minimal cost set of r arcs is found such that there is just one arc from each cluster and just one arc to each cluster. That is, if the clusters are thought of as vertices then an AP must be solved using the cost matrix $(\tilde{c}_{ij})$ where

$$\tilde{c}_{ii} = \infty \qquad \begin{cases} \tilde{c}_{ij} = c'_{ij} = \min_{\substack{\alpha \in V_i \\ \beta \in V_j}} c_{\alpha\beta}, & i \neq j \\ \tilde{c}_{ij} \leftarrow \min_k (c'_{ij}, c'_{kj}) & \text{if } c_{ij} > c_{it} + c_{tj} \text{ for some } t \end{cases}$$

If the solution is not a single circuit then the process may be repeated. Clearly, the worst situation arises when the AP solutions comprise circuits of length 2 (figure 7.5). However, Christofides has shown that even in this case the total effort, required to find a lower bound, is no more than 14 per cent greater than that required to solve the first AP (of size n).

Example 7.3   Find an improved lower bound for the TSP of example 7.1.

Solution   Using the reduced cost matrix (given in example 7.2) which results from first column then row reductions leads to the new AP with cost matrix

The Travelling Salesman and Chinese Postman Problems   157

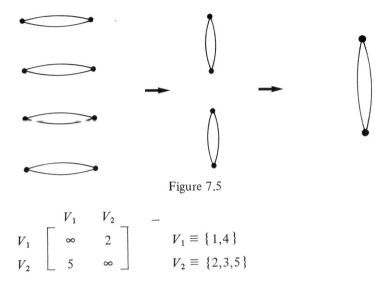

Figure 7.5

$$\begin{array}{c} & \begin{array}{cc} V_1 & V_2 \end{array} \\ \begin{array}{c} V_1 \\ V_2 \end{array} & \left[\begin{array}{cc} \infty & 2 \\ 5 & \infty \end{array} \right] \end{array} \qquad \begin{array}{c} V_1 \equiv \{1,4\} \\ V_2 \equiv \{2,3,5\} \end{array}$$

This clearly has an optimal solution of value 7 leading to the improved lower bound of $27 + 7 = 34$ for TSP. This is still short of the optimal value of 37, found in example 7.1, but represents a considerable improvement. It suggests that for larger problems improved bounds should be used if bounds based on AP are employed.

7.2 OTHER APPROACHES TO TSP

In order to simplify the discussion of the use of MSTs for solving travelling salesman problems, it is convenient first to transform TSP to the *open TSP* [denoted by OTSP(s,f)] in which a shortest Hamilton path is sought from vertex s to vertex f (s and f distinct).

Definition 7.2
A *Hamilton path (chain)* in a graph on n vertices is a path (chain) which visits each vertex just once and contains $n - 1$ arcs (edges).

Theorem 7.4
An n-vertex TSP can be solved by first solving an $(n+1)$-vertex OTSP(s,f).

Proof Suppose an optimal Hamilton circuit is sought in a network $N = (V,A,c)$ on n vertices which may without loss of generality be assumed to be complete (otherwise 'missing' arcs could be included with weight ∞). A new complete network $N^* = (V^*,A^*,c^*)$ is now formed by replacing any vertex $x \in V$ by two distinct vertices s and f. Formally

$$V^* = (V - \{x\}) \cup \{s,f\}$$

and c^* is defined by

$$\left.\begin{array}{ll} c^*_{ab} = c_{ab} & \text{all } a,b \notin \{s,f\} \\ c^*_{sb} = c_{xb} + M & b \neq f \\ c^*_{af} = c_{ax} + M & a \neq s \\ c^*_{sf} = c^*_{fs} = 2M & \\ c^*_{aa} = \infty & \text{all } a \in V^* \end{array}\right\} \quad (7.1)$$

where M is a very large number. (The reason for choosing to add the terms in M will be made clear later.)

Clearly, to each Hamilton circuit $\pi \equiv xu_1u_2\ldots u_{n-1}x$ in N there corresponds the unique Hamilton path $\pi^* \equiv su_1u_2\ldots u_{n-1}f$ in N^* and vice versa. Moreover the length of π^* with respect to c^* is just $2M$ plus the length of π with respect to c. The result follows immediately. □

Suppose now that we have a symmetric cost matrix (c^*_{ij}) with the form of equations 7.1, and use it to find an MST on N^*. Such a tree will contain n edges and, since M is very large, will contain

(1) a single edge sa, $a \neq f$, from s;
(2) a single edge bf, $b \neq s$, to f.

If also the tree has no vertex with degree in excess of 2 then it must be a Hamilton chain and hence since it is a minimal spanning tree it must be a shortest Hamilton chain (from s to f).

Example 7.4   Consider the symmetric cost matrix

$$\begin{bmatrix} \infty & 210 & 222 & 211 & 207 & 400 \\ & \infty & 31 & 33 & 37 & 210 \\ & & \infty & 34 & 20 & 222 \\ & & & \infty & 15 & 211 \\ & & & & \infty & 207 \\ & & & & & \infty \end{bmatrix}$$

with $M = 200$ added to rows and columns 1 and 6. Obtain a shortest Hamilton chain from vertex 1 to vertex 6.

Solution   Since a Hamilton chain from 1 to 6 cannot contain both 15 and 65 and since $c_{1j} = c_{6j}$ all j we may exclude one of these edges. We choose to exclude edge 15. The unique MST is shown in figure 7.6a.

In this case the MST fails to be a chain because vertex 5 has degree 3. However, since it is the optimal solution to a relaxed problem, the weight 483 of the MST must be a lower bound to the length of a shortest Hamilton chain.

The Travelling Salesman and Chinese Postman Problems

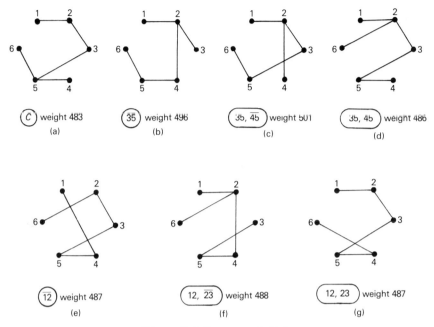

Figure 7.6 See example 7.4

Noting that only two of the edges 35, 45 and 65 can belong to a Hamilton chain we split the original problem into three subproblems

P_1 :	exclude 35
P_2 : include 35	exclude 45
P_3 : include 35 and 45	exclude 65

(For P_3, 15 and 25 may also be excluded for feasibility.) Exclusion of an edge ij can be achieved simply by replacing c_{ij} by ∞. Edges are included by starting with the set of included edges and then using algorithm (K) or algorithm (P) (section 2.3), suitably modified, to complete the tree.

The MSTs for the three subproblems are as shown in figures 7.6 b to d.

The MST for the subproblem P_3 resulting from the inclusion of edges 35 and 45, has least weight of the three, and so P_3 is split into three subproblems P_{31}, P_{32} and P_{33}. The MSTs corresponding to these new subproblems are shown in figures 7.6 e to g. The results are collected together by the B & B search tree of figure 7.7, from which it is seen that 145326 and 123546 are optimal solutions.

The method in the above example is essentially that of Christofides (1970). Its major defect is that the bounds are not sufficiently strong implying that the search tree will be undesirably large for larger problems. A way of improving the bounds is also obtained from Christofides (1970) and relies on the following analogue of theorem 7.1.

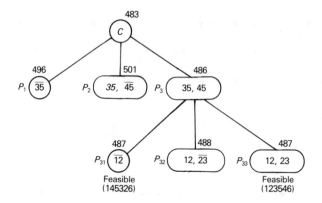

Figure 7.7 The B & B search tree for the problem of example 7.4

Theorem 7.5
The length of each Hamilton chain between s and f changes by the same amount $2(\sum_x p_x) - p_s - p_f$ if edge lengths are modified by

$$c'_{ij} = c_{ij} + p_i + p_j \tag{7.2}$$

The weights of spanning trees are not in general changed by the same amount by the transformation of equation 7.2. This usually allows a tighter bound to be obtained by choosing suitable values for the penalties p_i.

Example 7.5 The initial lower bound for the problem of example 7.4 was 483. By choosing suitable values for the penalties obtain a tighter bound.

Solution We note from figure 7.6 that vertex 5 is too popular (degree in excess of 2). Accordingly we penalise it by setting $p_5 = 12$ and $p_i = 0, i \neq 5$. The new matrix is

$$\begin{bmatrix} \infty & 210 & 222 & 211 & 219 & 400 \\ & \infty & 31 & 33 & 49 & 210 \\ & & \infty & 34 & 32 & 222 \\ & & & \infty & 27 & 211 \\ & & & & \infty & 219 \\ & & & & & \infty \end{bmatrix}$$

The minimal spanning tree is shown in figure 7.8.

The new initial lower bound is $510 - 2p_5 = 486$. Vertex 2 is now too popular and the bound can be improved by 1 to 487 if p_2 is set to 1. Edges 23, 35 and 45 are in each MST, and also 12 or 14 and 26 or 46 must be included. Thus from the 4 MSTs two, 123546 and 145326, are Hamilton chains and are hence optimal.

Using the above approach lower bounds can often be improved considerably and the greatest lower bound frequently obtained. However the maximum bound achievable using penalties may still fall short of the length of an optimal Hamilton chain giving rise to what is often termed a 'duality gap'.

Although it is not difficult to find suitable penalties for small problems this is not very easy for larger ones. Christofides (1970) offers some suggestions on how this might be achieved. Held and Karp (1971) used 'subgradient optimisation' which is guaranteed to converge to an optimal solution though the convergence may be fairly slow. Some aspects of Held and Karp's method were improved upon by Hansen and Krarup (1974).

Currently, asymmetric TSPs with n over 200 and symmetric TSPs with n over 100 are being solved exactly. For more details and for further developments the reader is referred to Padberg and Hong (1980).

The state of the art of finding exact solutions to TSPs is very highly developed. However, TSPs are difficult to solve and consequently for very large problems one should expect to have to resort to heuristic methods, to avoid using an inordinate amount of computing resources.

A well-known and very successful heuristic method is that due to Lin (1965). For any tour π, select an ordered set $Q = \{ab, pq, uv\}$ of three of its edges (arcs) and remove them. A new tour $T^Q_j(\pi)$ is now formed by adding three edges (arcs) at least one of which was not among those removed. The 7 possibilities for $T^Q_j(\pi)$ for an undirected graph are shown in table 7.1.

$T^Q_j(\pi)$ will be termed a *3-neighbour* of π and T^Q_j a *3-transformation*. Notice that if any of the part tours from b to p, q to u or v to a consists of a single vertex then not all of the $T^Q_j(\pi)$ will be distinct. Also if we are considering directed tours then only $T^Q_4(\pi)$ is allowable.

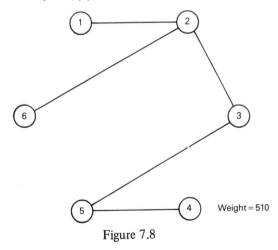

Figure 7.8

Table 7.1

Transformation	Edges replacing $Q = \{ab, pq, uv\}$		
T^Q_1	ab	pu	qv
T^Q_2	ap	bq	uv
T^Q_3	ap	bu	qv
T^Q_4	aq	ub	pv
T^Q_5	aq	up	bv
T^Q_6	au	qb	pv
T^Q_7	au	qp	bv

If $\varphi(\pi)$ denotes the length of a tour π then in Lin's method π is replaced, if possible, by some 3-neighbour $T^Q_j(\pi)$ for which $\varphi(T^Q_j(\pi)) < \varphi(\pi)$. The 3-neighbours of this new tour are searched for one which gives an improvement (that is, reduces total tour length). This is continued for as long as possible giving a sequence of tours

$$\pi = \pi_0, \pi_1, \pi_2, \ldots, \pi_r = \tilde{\pi}$$

such that

(1) $\varphi(\pi_i) < \varphi(\pi_{i-1})$ $\quad i = 1, \ldots, r$
(2) for all 3-neighbours π' of $\tilde{\pi}$,

$$\varphi(\tilde{\pi}) \leq \varphi(\pi')$$

That is, starting with a 'trial' tour π, small adjustments are repeatedly made leading to successively better and better tours until a tour $\tilde{\pi}$ is obtained which cannot be improved by such a small adjustment. $\tilde{\pi}$ is called a *local optimum* since it is at least as good as any of its 'neighbouring' tours.

Example 7.6 Given the reduced cost matrix of example 7.2, and starting with the trial tour $\pi \equiv 123451$, use Lin's method to find a solution locally optimal with respect to 3-transformations.

Solution The weights of the arcs 12, 23, 34, 45 and 51 are 2, 0, 8, 6 and 16 respectively and so it seems sensible to try $Q = \{34, 45, 51\}$. Now

$$T^Q_4(\pi) = 123541$$

and has length 13 which is less than that of π, and consequently π_1 is set to 123541. There are $\binom{5}{3} = 10$ possible choices for Q namely

$Q_1 = \{12, 23, 35\}$ $\quad Q_6 = \{12, 54, 41\}$
$Q_2 = \{12, 23, 54\}$ $\quad Q_7 = \{23, 35, 54\}$
$Q_3 = \{12, 23, 41\}$ $\quad Q_8 = \{23, 35, 41\}$
$Q_4 = \{12, 35, 54\}$ $\quad Q_9 = \{23, 54, 41\}$
$Q_5 = \{12, 35, 41\}$ $\quad Q_{10} = \{35, 54, 41\}$

Denoting $T^{Q i_4}(\pi)$ simply by T_i we have

T_1 = 132541 length 14
T_2 = 135241 length 25
T_3 = 135421 length 27
T_4 = 152341 length 27
T_5 = 154231 length 27
T_6 = 142351 length 24
T_7 = 125341 length 10
T_8 = 125431 length 33
T_9 = 124351 length 44
T_{10} = 123451 length 32   (trial solution at previous step)

It is found that the best of these (T_7) cannot be further improved by a 3-transformation and so is locally optimal. In fact it is globally optimal as we have seen earlier.

More generally we may consider λ-transformations, in which λ arcs (or edges) are removed from a tour and replaced by another set of λ arcs (or edges) so as to form a tour. This leads to λ-*opt* tours being generated. Experimental experience has shown that for λ odd the quality of λ-opt tours is comparable to that of (λ+1)-opt tours but the number of neighbours of a given tour is much less. Hence it seems not to be worthwhile taking λ = 4,6, ... Also λ = 5 leads to much searching of neighbours and it is probably better to take λ = 3 and perform more hill-climbs. This leaves the possibility of λ = 2 for the symmetric case which may be useful for finding trial solutions for a λ = 3 hill-climb (see section 11.1).

An extension of Lin's method is that due to Lin and Kernighan (1973). Starting with a fixed k and trial solution π_0 consider the sequence $\pi_0, \pi_1, \ldots, \pi_k$ with π_i obtained from π_{i-1} by a λ-transformation producing a maximum decrease (which may be negative) in tour length. If $g = \max_{j \leq k} \{\varphi(\pi_0) - \varphi(\pi_j)\}$ is greater than zero with the maximum occuring for $j = j^*$ then the process is repeated with π_{j*} replacing π_0; otherwise the process is terminated.

7.3 THE CHINESE POSTMAN PROBLEM AND MATCHING

The preceding sections have dealt with finding minimal length Hamilton circuits (or cycles). A similar problem is that of finding a minimal length circuit (or cycle) which passes along each arc (or edge) of a network at least once. In this section the discussion will be restricted to undirected graphs and for this case the routing problem just mentioned is often referred to as the Chinese postman problem (Kwan, 1962) and will be abbreviated to CPP. Such problems are of interest, for example, regarding refuse collection, milk or post delivery, salting and gritting of roads to prevent ice formation and inspection of electric power and other networks (Christofides, 1975). If an Euler cycle exists then clearly this provides a solution; otherwise some edges must be traversed more than once.

For a network $N = (V,E,w)$ let V_O denote the set of vertices with odd degree and $V_E = V - V_O$ denote the set of vertices of even degree. It is easily proved that $|V_O|$ is even.

Definition 7.3
A *matching set of chains* on N consists of a set C of $|V_O|/2$ chains between vertices of V_O such that each $\xi \in V_O$ is at the end of precisely one of these chains.

Theorem 7.6
Define $c(C) = \sum_{\gamma \in C} \sum_{ij \in \gamma} c_{ij}$ to be the combined cost of chains in C. Then a matching set of chains C for which $c(C)$ is minimal is such that no two chains of C possess a common link provided $c_{ij} > 0$ for all i,j.

Proof This is left as an exercise for the reader.

Theorem 7.7
Let a cycle σ be a feasible solution to CPP over a network $N = (V,E,c)$ for which $c_{ij} > 0$ all i,j. Then σ includes one copy of every edge of N together with a matching set of chains C.

Proof Let H_r be the hypothesis that the desired result is true if $|V_O| = 2r$. Clearly H_0 is true. Assume now that H_r is true and that $|V_O| = 2r + 2$. It is clear that σ must contain each edge at least once. Let μ denote the remaining edges.

Since σ is a cycle a traversal of σ leads to each vertex being 'entered' and 'left' an equal number of times and so σ must contain an even number of edges incident with each $x \in V_O$. Hence, some edge of N incident with x, xa say, must be repeated in σ. If $a \in V_E$ then, since xa is repeated, there must be another repeated edge, ab say, which is also incident with a. If $b \in V_E$ then there will be a repeated edge bc, etc. Thus a chain $xabc \ldots gy$ is constructed until the first $y \in V_O$ is encountered. Hence μ contains edges which form a chain between x and y.

Now let N^* be the 'network' N with one copy of each of the edges xa, ab, $bc, \ldots, gy$ added. (Note that (V^*, E^*) now has repeated edges and so is a multigraph rather than a graph — see exercise 2.3.) σ is a solution to CPP over N^* and $|V^*_O| = 2r$, and so σ contains all the edges of N^* together with a matching set of chains (on N^*). But this implies immediately that σ contains all the edges of N once each together with a matching set of chains. $\square$

Theorem 7.8
Let $N = (V,E,c)$ be a network with $c_{ij} > 0$ all $ij \in E$. Then E together with the edges of a minimal cost matching set of chains provides an optimal solution to CPP on N.

Proof Let C^* be a matching set of chains of minimal cost. Then the cost of an optimal solution to CPP cannot, since $c_{ij} > 0$, be less than $\sum_{ij \in E} c_{ij} + c(C^*)$ Since the set E together with the edges of C^* forms a feasible solution to CPP the result follows. □

Example 7.7 A milkman and his assistant deliver to houses in each of the streets of a housing estate whose plan is given in figure 7.9a. At the start of his round he travels from the dairy to point A and finally returns to the dairy via A. Milk is delivered to both sides of each street at the same time, and the route followed is $AYZFYXZEBEDFDCAXBCA$. Can this route be improved?

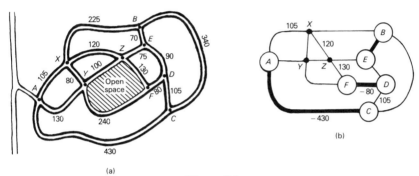

Figure 7.9

Solution Figure 7.9b shows the graph corresponding to this problem. The vertices of odd degree are A, B, C, D, E and F. The repeated edges are AC, BE and DF which together form a matching set of chains C. The milkman will travel along the edges BE, DF and AC once each without delivering, and the given route leads to least travelling without delivery if it is an optimal solution to CPP, that is, if C is a minimal cost matching set of chains.

Consider now the network N' derived from N by negating the costs on AC, BE and DF (figure 7.9b). Cycle $AXZFDCA$ has cost $105 + 120 + 130 - 80 + 105 - 430 = -50 < 0$. Replace the repeated edges AC, DF on this cycle by edges AX, XZ, ZF and DC. This leads to a matching set of chains

$$C' = \{AXZF, CD, BE\}$$

with cost $\quad c(C') = (105 + 120 + 130) + 105 + 70 = 530$
$\qquad\qquad\quad = c(C) - 50$

Let N'' be N with the costs of AX, XZ, ZF, CD and BE negated. There are no negative length cycles in N'' and we conclude that C' gives an optimal solution to CPP.

The improvement (from C to C') was small and the route defined by C may be thought preferable if all repeated streets with houses on both sides have delivery one side only on each journey down those streets.

The above example introduces the ideas of the following, easily proved, theorem.

Theorem 7.9
A matching set of chains C has minimal cost if and only if the network N', obtained from N by negating the costs of edges in C, possesses no negative cost elementary cycle.

If a negative length elementary cycle σ is present then an improved matching set of chains C' is obtained by removing edges which are in both C and σ and adding edges in σ *but not in* C.

The above approach, due to Kwan (1962) is conceptually straightforward but requires an algorithm for the detection of negative length cycles.

In many cities the majority of road intersections are the junction of four roads and so, in the corresponding model network, the proportion of vertices with odd degree is fairly small. Thus it may be preferable first to find shortest chains between every pair of distinct vertices in V_O, and then to find an optimal matching set of chains by obtaining a minimal cost pairing (or *matching*) of vertices in a graph with vertex set V_O.

Example 7.8 Solve CPP for the problem of example 7.7 using the modified approach outlined above.

Solution The least cost (shortest distance) matrix between the vertices in V_O is

	A	B	C	D	E	F
A	∞	330	430	390	300	355
B	330	∞	265	160	70	240
C	430	265	∞	105	195	185
D	390	160	105	∞	90	80
E	300	70	195	90	∞	170
F	355	240	185	80	170	∞

There are 15 distinct matchings $\{AB,CD,EF\}, \{AB,CE,DF\}$, etc., of which the least cost is $\{AF, BE, CD\}$ with cost 530. Noting that AF corresponded to the chain $AXZF$ in the original network this leads us to the optimal matching set of chains $C' = \{AXZF, BE, CD\}$ as before.

For this small problem complete enumeration was used but this would not be practicable for larger problems. (The number of distinct matchings is $(2r-1) \cdot (2r-3) \cdot \ldots \cdot 3 \cdot 1 = (2r)!/r!\, 2^r$ where $2r = |V_O|$.)

A method for obtaining minimal cost complete matchings, which is guaranteed to run in polynomial time, has been devised by Edmonds. Finding an optimal matching in a bipartite graph is just an assignment problem but the extension to more general networks is beyond the scope of the present text and the reader is referred to Lawler (1976) and references therein. However, the simpler problem of finding a maximum cardinality matching, will be treated briefly.

Definition 7.4
A matching M^* of a network N is a *maximal cardinality matching* if no other matching M of N contains more edges; that is, $|M^*| \geq |M|$.

Definition 7.5
If M is a matching of network N over vertex set V then $x \in V$ is *covered* if x is an end point of an edge in M. A vertex is *exposed* if it is not covered.

Let b be an exposed vertex then an *M-alternating tree* (X, E) is constructed using the following algorithm.

Algorithm (AT)
(Generating alternating trees)

Step 1 (Setup)
    Set $l(b) = (+)$ for some exposed vertex b.
    Set $E = \phi$, $X = \{b\}$ and $\text{scan}(y) = \textit{false}$ for all y.

Step 2 (Termination)
    Select a vertex $x \in X$ for which $\text{scan}(x) = \textit{false}$.
    (If this is impossible then terminate.)

Step 3 (Iteration)
    If $l(x) = (+)$ then perform step 3a else step 3b.

    Step 3a $[l(x) = (+)]$
        For all edges $xu \notin M$ with $u \in X$
        add u to X and xu to E, and set $l(u) = (-)$.

    Step 3b $[l(x) = (-)]$
        If there is an edge $xv \in M$ then add v to X,
        add xv to E and set $l(v) = (+)$.
    Set $\text{scan}(x) = \textit{true}$ and return to step 2.

If at any stage an exposed vertex $c \neq b$ is labelled (and the label must be $(-)$, why?) then there is a chain $\pi = bv \ldots sc$ in the alternating tree, from b to c, with alternate edges not in and in M. Since neither bv nor sc is in M, an improved matching M' can be obtained from M by removing from M those edges which are also in π and adding to M the other edges of π. Clearly $|M'| = |M| + 1$.

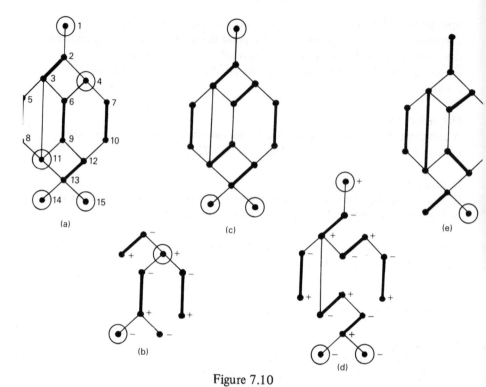

Figure 7.10

Example 7.9 Find a maximal cardinality matching in the graph of figure 7.10a from the matching M shown.

Solution

Step
1. Vertex 4 is exposed.
   $l(4) = (+), \quad X = \{4\}, \quad E = \phi$
2. Vertex 4 is on the tree and unscanned (that is, scan(4) = *false*).
3a. $X \leftarrow \{4,2,6,7\}, \quad E \leftarrow \{42, 46, 47\},$
    $l(2) = l(6) = l(7) = (-), \quad \text{scan}(4) = true.$
2. Select vertex 6.
3b. $X \leftarrow \{4,2,6,7,9\}, \quad E \leftarrow \{42, 46, 47, 69\},$
    $l(9) = (+), \quad \text{scan}(6) = true.$
2. Select vertex 2.
3b. $X \leftarrow \{4,2,6,7,9,3\}, \quad E \leftarrow \{42, 46, 47, 69, 23\}.$
    $l(3) = (+), \quad \text{scan}(2) = true.$
2. Select vertex 7.
3b. $X \leftarrow \{4,2,6,7,9,3,10\}, \quad E \leftarrow \{42, 46, 47, 69, 23, 7\text{--}10\}.$
    $l(10) = (+), \quad \text{scan}(7) = true.$

2 Select vertex 9.
3a $X \leftarrow \{4,2,6,7,9,3,10,11,12\}$,
    $E \leftarrow \{42, 46, 47, 69, 23, 7\text{--}10, 9\text{--}11, 9\text{--}12\}$,
    $l(11) = l(12) = (-)$.

Since the chain 4–6–9–11 has been found between exposed vertices 4 and 11 a new and improved matching

$$M' = (M - \{69\}) \cup \{46\} \cup \{9\text{--}11\}$$

shown in figure 7.10c is obtained. A new alternating tree is now grown from exposed vertex 1 (figure 7.10d) and an improved matching M'' obtained (figure 7.10e). This last clearly has maximal cardinality.

The above example illustrates the following result the proof of which is left as an exercise.

Theorem 7.10
A matching M has maximal cardinality if and only if no alternating tree can be grown which contains a chain between two oppositely labelled exposed vertices.

The approach can be modified to apply also to non-bipartite graphs (see for example, Christofides, 1975).

EXERCISES

7.1 π is a minimal length cycle in a network N visiting every vertex at least once. Prove that π is an optimal solution to TSP if the 'triangle property' is satisfied; that is, $c_{ij} + c_{jk} \geqslant c_{ik}$ for all $i \neq j \neq k \neq i$.

7.2 An assignment problem with vertex set $\{1, 2, \ldots, n\}$ has (reduced) cost matrix (c_{ij}) such that all elements are non-negative and there is at least one zero element in each row and each column. Let $G = (X \cup Y, E)$ be the undirected bipartite graph for which

$$X = \{1, 2, \ldots, n\}, \quad Y = \{\bar{1}, \bar{2}, \ldots, \bar{n}\}, \quad E = \{i\bar{j} \mid c_{ij} = 0\}$$

and let M be a maximum cardinality matching. Verify that if $|M| = n$ then this defines a zero cost, and hence optimal, solution to AP.

7.3 Suppose M (as defined in exercise 7.2) is such that $|M| < n$, then there must be an exposed vertex $\bar{j} \in Y$. Let T be the set of vertices which can be reached from $\bar{j}$ by M-alternating chains. Prove that

(a) $\bar{j}$ is the only exposed vertex in T;
(b) $|T \cap Y| = |T \cap X| + 1$;

(c) for some $\epsilon > 0$ there is an admissible transformation $c'_{ij} = c_{ij} - p_i - q_j$, $1 \leq i,j \leq n$ where $p_i(q_j)$ is equal to $\epsilon(-\epsilon)$ if $i \in T \cap X (j \in T \cap Y)$ and is zero otherwise.

Deduce the result of theorem 7.3.

7.4 It follows from exercises 7.2 and 7.3 that an AP may be solved by the algorithm below (which is a variant of the well-known Hungarian method).

Algorithm (HM)
(To solve an AP)

Step 1 (Setup)
Reduce the cost matrix by an admissible transformation so that there is a zero element in each row and column.

Step 2 (Termination)
Find a maximum cardinality matching M of G.
If $|M| = n$ terminate; M determines the optimal assignments.

Step 3 (Updating cost matrix)
Update cost matrix as in exercise 7.3(c) with

$$\epsilon = \min_{j \notin T \cap Y} [\min_{i \in T \cap X} (c_{ij})]$$

Further reduce any column with no zero element.
Go to step 2.

Solve the AP whose cost matrix is

$$\begin{bmatrix} \infty & 15 & 24 & 7 & 10 & 20 \\ 3 & \infty & 0 & 0 & 0 & -2 \\ 39 & 5 & \infty & 16 & 19 & 19 \\ 23 & 13 & 24 & \infty & 12 & 4 \\ 9 & 5 & 9 & 0 & \infty & 6 \\ 23 & 7 & 19 & -1 & 18 & \infty \end{bmatrix}$$

By contracting subtours and solving a further AP (as in example 7.3) find an improved bound for the TSP with the above cost matrix.

7.5 Solve the TSP with the cost matrix of exercise 7.4 using
(a) Little's method;
(b) the Eastman method.

7.6 Problem P requires a minimal length Hamilton path to be found in a complete network $N = (X, A, c)$. A new network $\tilde{N} = (X \cup \{u\}, A \cup \bar{A}, \tilde{c})$ is formed from N by adding a vertex u and arcs in the set $\bar{A} = \{ux \mid x \in X\} \cup \{xu \mid x \in X\}$, and defining

$$\tilde{c}_{xy} = c_{xy} \quad \text{all } x, y \in X$$
$$\tilde{c}_{uu} = \infty$$
$$\tilde{c}_{ux} = \tilde{c}_{xu} = 0 \quad \text{all } x \in X$$

Prove that $x_1 x_2 \ldots x_n$ is an optimal solution to problem P if, and only if, $u x_1 x_2 \ldots x_n u$ is an optimal solution to TSP on $\tilde{N}$.

Using this correspondence between problems find an optimal solution to the sequencing problem of example 2.7.

7.7 Prove that for a symmetric TSP any three distinct cost matrix elements c_{ij}, c_{jk} and c_{ki} may be set to ∞ without eliminating all minimal length tours.

7.8 Reddi and Ramamoorthy (1972) considered the flow shop sequencing problem (exercise 3.6) with the extra constraint that no job should have to wait while a previous job is still being processed on machines $2, 3, \ldots, m$. That is, no in-process inventory is allowed. Given the data of exercise 3.6, a schedule for the sequence J_1, J_2, J_3, J_4 is shown in figure 7.11. If $m = 3$ and J_q immediately follows J_p, then J_q must wait a time $F(p, q)$ after processing of J_p on M_1 is complete before processing of J_q may begin. Verify that

$$F(p, q) = \max\,[0, t_{2p} - t_{1q}, (t_{2p} + t_{3p}) - (t_{1q} + t_{2q})]$$

Suppose that for sequence $J_1, J_2, \ldots, J_n$ job J_i cannot start on M_1 before T_i and that $T_1 = 0$, then

$$T_1 = 0$$
$$T_2 = T_1 + t_{1j_1} + F(j_1, j_2)$$
$$\vdots$$
$$T_{i+1} = T_i + t_{1j_i} + F(j_i, j_{i+1})$$
$$\vdots$$
$$T_n = T_{n-1} + t_{1j_{n-1}} + F(j_{n-1}, j_n)$$

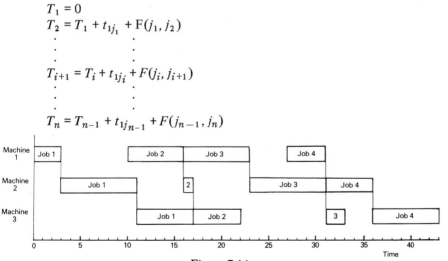

Figure 7.11

and the completion time T_c is given by

$$T_c = \sum_{i=1}^{n} t_{1j_i} + \sum_{i=1}^{n-1} F(j_i, j_{i+1}) + \sum_{i=2}^{m} t_{ij_n}$$

Prove that a minimal makespan sequence of jobs is provided by an optimal solution to the TSP with vertex set $\{0, 1, \ldots, n\}$ and cost matrix (c_{ij}) where

$$c_{ij} = F(i, j) \qquad 0 \neq i \neq j \neq 0$$

$$c_{0j} = \sum_{j=1}^{n} t_{1j}$$

$$c_{j0} = \sum_{i=2}^{m} t_{ij} \qquad c_{ii} = \infty \text{ for all } i$$

Solve the no in-process inventory machine sequencing problem for which $m = 3$, $n = 4$ and processing times are as given in exercise 3.6.

8 Distribution Problems

The provision of goods or services from a set of supply points to a set of demand points is of great economic and social importance. For example; raw materials need to be supplied from various sources to sites of manufacturing; finished goods are transported from factory to depots or supplied direct to customers; children are transported to and from school by bus; customer orders and enquiries 'travel with' a commercial traveller from actual and potential customers back to the traveller's base; fire engines travel from fire stations to the scene of fires, etc. Clearly the question of an efficient distribution system is closely tied to locational decisions (siting of factories, warehouses, emergency facilities, etc.). In this chapter the locations of facilities within a transportation system will be assumed to be fixed and given, and the routing decisions needed to provide an efficient distribution system will be explored.

Each vehicle (train, lorry, ambulance, etc.) will have a capacity and the way in which this capacity relates to the volume of goods to be transported may be used to categorise a distribution system. The shipment of goods to depots is often carried out in bulk whereas supplying customers usually involves smaller consignments. Let $\{s_i\}_{i \in I}$ be a set of sources, $\{d_j\}_{j \in J}$ a set of destinations, t_{ij} the quantity of goods to be carried, per unit time, from s_i to d_j, and finally let c be the capacity of a typical vehicle. If $c < t_{ij}$, as may occur for bulk delivery, then capacity implications are less demanding and routing decisions may be made with regard to shortest routes (chapter 4) or maximal, or least cost, flow considerations (chapters 9, 10). If $c = \infty$ (that is, capacity constraints may be ignored) then, in the absence of other constraints, one vehicle suffices to supply goods to all destinations and the problem can be modelled as a TSP (chapter 7). This chapter will be concerned with the intermediate situation $t_{ij} < c < \infty$ in which a single vehicle may be used to supply several customers.

Other constraints will often be operative in practice, for example, maximum travel time or distance allowed per vehicle, due times for deliveries, permitted delivery intervals and loading considerations. Some of these are touched on in subsequent pages but the main emphasis will concern capacity constraints.

From now on supply points will be termed depots and destinations will be termed customers. Problems in which there is a single depot, that is $|I| = 1$, (and by symmetry, problems with a single destination) will be considered in section 8.1. The discussion of single depot problems is continued in section 8.2, and in section 8.3 multi-depot problems are considered briefly.

8.1 SINGLE-DEPOT VEHICLE ROUTING PROBLEMS: TSP AND SAVINGS BASED METHODS

The simplest form of vehicle routing problem (VRP), and one which has been the subject of considerable investigation, concerns a single depot from which n customers must be supplied by a fleet of vehicles where

(1) each customer i has a known location and a fixed requirement q_i which must be met;
(2) all vehicles are the same and have a capacity Q (representing some combination of weight, volume, etc.) which must not be exceeded.

The objective is to determine routes for the vehicles so that

O1: the total travel cost (distance, time, etc.) is minimal, or
O2: the required fleet size is minimal.

If Q is large compared to the average value of the q_i, and objective O1 is of interest then, we may expect the basic VRP to have characteristics in common with TSP (which problem it becomes as $Q \to \infty$). This observation, together with the likelihood that the costs are approximately Euclidean, prompts the following results for TSP, all of which are easily proved.

Theorem 8.1
If the 'triangle property'

$$c_{ik} \leq c_{ij} + c_{jk} \qquad i,j,k = 1,2,\ldots,n \tag{8.1}$$

holds then there is a minimal length circuit visiting every vertex i which is also a tour (visits each vertex no more than once).

Theorem 8.2
If the triangle property (inequality 8.1) is satisfied then there is a minimal length tour which has no 'crossings' (figure 8.1).

Theorem 8.3
Every optimal tour of a Euclidean TSP visits the vertices which are on the convex hull in either clockwise or anticlockwise order (see figure 8.2).

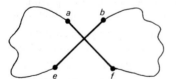 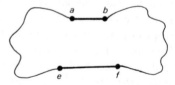

Figure 8.1 The tour on the right (with the crossing removed) is shorter than the tour on the left

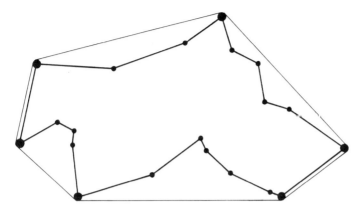

Figure 8.2 The convex hull is indicated by the enclosing hexagon

Returning now to the basic VRP let V_k be the *call set* of the kth vehicle, that is, the set of customers visited by the kth vehicle plus the depot (vertex 0). Then if the route of this vehicle is part of an optimal solution it must form an optimal tour on V_k. The above results (theorems 8.1 to 8.3) are then applicable to routing the vehicle if the set V_k is prescribed. It follows that the combination of all the routes comprising an optimal solution has a 'petal like' appearance, though it is possible that two routes might overlap (figure 8.10) because of the capacity constraint.

How are the sets V_k to be determined? The TSP connection can be further exploited as follows. Split vertex 0 (the depot) into vertices named $-N, -N+1, \ldots, -2, -1$ (called *pseudo depots*) and define a new cost matrix $(\tilde{c}_{ij})$ by

$$\begin{aligned}
\tilde{c}_{ij} &= c_{ij} & i,j &= 1,2,\ldots,n \\
\tilde{c}_{\alpha j} &= c_{0j} & \alpha &= -1,-2,\ldots,-N, \quad j = 1,2,\ldots,n \\
\tilde{c}_{i\beta} &= c_{i0} & i &= 1,2,\ldots,n, \quad \beta = -1,-2,\ldots,-N \\
\tilde{c}_{\alpha\alpha} &= \infty & \alpha &= -1,-2,\ldots,-N \\
\tilde{c}_{\alpha\beta} &= \mu & \alpha,\beta &= -1,-2,\ldots,-N \text{ and } \alpha \neq \beta
\end{aligned}$$

Any tour π relative to the cost matrix $(\tilde{c}_{ij})$ consists of visits to customers (one to each) interspersed by periodic visits to a pseudo depot. Let $\alpha i_1 i_2 \ldots i_p \beta$ be part of π between successive pseudo depots α and β then a corresponding vehicle route on call set $V = \{i_1, i_2, \ldots, i_p, 0\}$ is

$$\rho: \text{depot} - i_1 - i_2 \ldots - i_p - \text{depot}$$

Since there are N pseudo depots and N call sets V_k, N vehicle routes will be defined. Thus a tour π corresponds to a potential set of vehicle routes $\{\rho_k\}$ (that is, to a solution to VRP) and vice versa.

The cost of π from α to β is

$$\tilde{c}_{\alpha i_1} + \tilde{c}_{i_1 i_2} + \ldots + \tilde{c}_{i_p \beta} = \begin{cases} c_{0i_1} + c_{i_1 i_2} + \ldots + c_{i_p 0} & \text{if } p > 0 \\ c_{\alpha\beta} = \mu & \text{if } p = 0 \end{cases}$$

leading to the following.

Theorem 8.4

If the cost of keeping a vehicle at the depot is μ then the cost of a solution to basic VRP relative to cost matrix (c_{ij}) is the same as the cost of the corresponding solution to TSP relative to cost matrix $(\tilde{c}_{ij})$.

Corollary (Christofides, 1976)   An optimal solution to the TSP on $n+N$ vertices relative to cost matrix $(\tilde{c}_{ij})$ and subject to the additional constraint

$$\sum_{j \in V_k - \{0\}} q_j \leq Q \qquad k = 1, \ldots, N$$

yields a solution to basic VRP which

(1) is the cheapest solution that uses *all* vehicles if $\mu > M$;
(2) is the cheapest solution (objective O1) if $\mu = 0$;
(3) is the cheapest solution subject to a saving in fixed cost of $-\mu$ per unused vehicle if $\mu < 0$;
(4) uses a minimal sized fleet in the cheapest way possible (objective O2) if $-\infty < \mu < -M$,

where M is a very large number

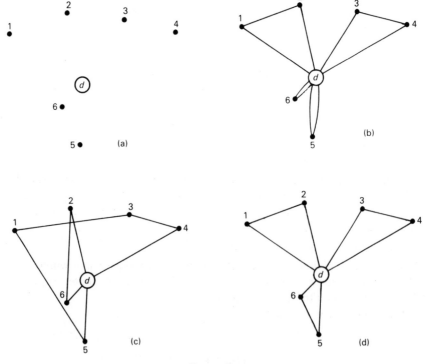

Figure 8.3

Distribution Problems

Example 8.1 Six customers with requirements q_k, $k=1,2,\ldots,6$ are to be supplied from a single depot (see figure 8.3a) by a fleet of four vehicles each of capacity 15 tons. If the cost matrix (c_{ij}) and requirements vector are as given below determine

(1) the cheapest solution which makes use of all four vehicles;
(2) the minimal fleet size and the cheapest solution making use of a fleet this size;
(3) the fleet size and routes leading to an overall minimal cost solution

$$(c_{ij}) = \begin{array}{c|ccccccc} & 0 & 1 & 2 & 3 & 4 & 5 & 6 \\ \hline 0 & \infty & 24 & 19 & 20 & 27 & 16 & 12 \\ 1 & 24 & \infty & 17 & 31 & 44 & 36 & 23 \\ 2 & 19 & 17 & \infty & 16 & 29 & 35 & 25 \\ 3 & 20 & 31 & 16 & \infty & 15 & 34 & 28 \\ 4 & 27 & 44 & 29 & 15 & \infty & 40 & 37 \\ 5 & 16 & 36 & 35 & 34 & 40 & \infty & 11 \\ 6 & 12 & 23 & 25 & 28 & 37 & 11 & \infty \end{array} \quad q = \begin{array}{c} 4 \\ 10 \\ 4 \\ 3 \\ 3.5 \\ 5 \end{array}$$

Solution (1) Following the construction of the matrix $(\bar{c}_{ij})$ described above gives (where d_i replaces $-i$)

	d_1	d_2	d_3	d_4	1	2	3	4	5	6
d_1	∞	∞	∞	∞	24	19	20	27	⑯	12
d_2	∞	∞	∞	∞	24	⑲	20	27	⑯	⑫
d_3	∞	∞	∞	∞	24	19	20	27	⑯	⑫
d_4	∞	∞	∞	∞	24	19	20	27	16	⑫
1	24	24	24	24	∞	17	31	44	36	23
2	19	19	⑲	19	17	∞	16	29	35	25
3	20	20	20	20	31	16	∞	15	34	28
4	27	27	27	27	44	29	15	∞	40	37
5	⑯	16	⑯	⑯	36	35	34	40	∞	⑪
6	⑫	⑫	⑫	12	23	25	28	37	11	∞

Since the problem is symmetrical any three arcs ij, jk and ki may be excluded (see exercise 7.7). We choose $d_1 5$, 56 and $6d_1$. Also since the pseudo depots are indistinguishable we also exclude $d_2 2$, $d_2 5$, $d_2 6$, $d_3 5$, $d_3 6$, $d_4 6$, $2d_3$, $5d_1$, $5d_3$, $5d_4$, $6d_2$, $6d_3$ (exercise 8.1). That is, all the circled elements of the above matrix may be set to ∞. The resulting matrix is then reduced by subtracting 12,20,19,16,17,12,13,15,11 and 11 from rows d_1 to 6 respectively and 7,5,7,1,4,0,0,2,0,0, from columns d_1 to 6 respectively giving a total reduction of 172. Relative to the reduced matrix there is a zero cost assignment solution $d_1 6 d_4 5 d_2 1 d_3 2 d_1$; 343. This is clearly not a feasible tour. It is convenient to break the subtour 343 by branching on arc 34, as would be done in Little's method (section 7.1). Performing maximal reductions gives zero cost assignments which are both tours. Since the bound in each case is 178 the corresponding solutions to VRP are optimal. These solutions are

depot–1–2–depot
depot–3–4–depot
depot–5–depot
depot–6–depot

and one which differs only in that the second route is reversed (figure 8.3b).
(2) Clearly fewer than $1/Q \, \Sigma q_k = 29/15$, vehicles will not suffice; that is, at least two vehicles are required. Moreover, there is only one way of partitioning $\{q_k\}$ into two groups so that the sum in neither group exceeds 15 tons, namely $\{2,6\}$ and $\{1,3,4,5\}$. By solving two straightforward TSPs it is found that an optimal solution to VRP with $N = 2$ is (figure 8.3c)

depot–2–6–depot
depot–5–1–3–4–depot

at a total cost of 181.

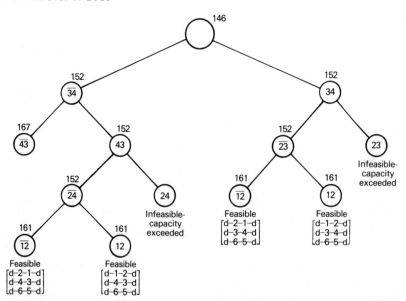

Figure 8.4 The B & B search tree for the problem of example 8.1 part (3)

Distribution Problems

(3) The least cost solution to VRP could be found by using part (2) of the corollary to theorem 8.4. However, VRP has been solved for $N = 4$ and $N = 2$ leaving $N = 3$ as the only other possibility. It is simpler to solve for $N = 3$ and compare with solutions found earlier. Arcs $d_1 5, d_2 5, d_2 6, d_3 6, 5d_1, 5d_3, 56, 6d_1, 6d_2$ can be excluded without losing optimality. The course of the solution is shown by the B & B tree in figure 8.4 in which all bounds correspond to maximal reduction.

Notice that if arc 34 is included in the tour then arcs 23 and 42 can be excluded since $q_2 + q_3 + q_4 > 15$.

An optimal solution to VRP is (figure 8.3d).

depot–1–2–depot
depot–6–5–depot
depot–3–4–depot

with cost 161. Since this is less than the costs of the cheapest solutions with $N = 2$ and $N = 4$ it is the overall cheapest solution.

According to Eilon et al. (1971), 'the computational efficiency of the branch-and-bound algorithm, when applied to the vehicle scheduling problem, is sub-

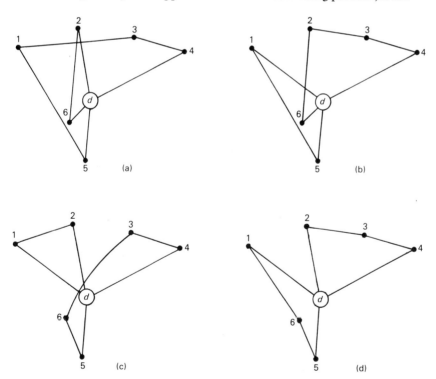

Figure 8.5 The solutions Φ, C, P and Y (see figure 8.6) are shown in (a), (b), (c) and (d) respectively

stantially reduced when compared with its efficiency in solving an equivalent travelling salesman problem'. This suggests the possibility of using an approximate TSP method; Lin's algorithm (section 7.2) has proved useful for this purpose (Christofides and Eilon, 1969).

Example 8.2 Given the data of example 8.1 try to find a solution better than that of figure 8.3c (shown again as figure 8.5a), if the possibility of replacing one of the vehicles by a larger one of capacity 20 tons is being considered.

Solution The starting tour is

π : depot–2–6–depot–5–1–3–4–depot

The aim is to improve π by successive allowable transformations (compare section 11.1). A common procedure is to use 2-transformations until a 2-opt capacity feasible tour is found, then continue with 3-transformations. We modify this slightly by considering only 2-transformations that produce a new tour and which remove crossings as defined by pictures such as those of figures 8.5a to d. (Since the graph is only approximately Euclidean, we consider arc 15 effectively to give rise to crossings with 6d and 26.)

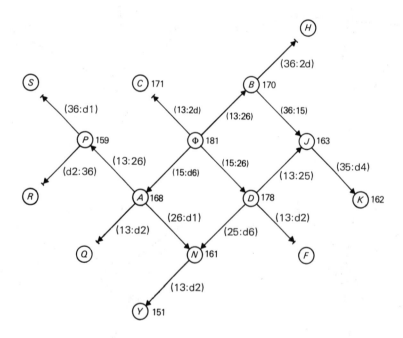

Figure 8.6 Possible solutions to VRP obtained from Φ by a series of crossing removals. Where an infeasible solution would be obtained a bar is placed at the arrowhead of the transformation. (See example 8.2)

A crossing of a pair of edges ij and kl will be denoted by $(ij:kl)$. π has four crossings $(15:d6)$, $(13:26)$, $(13:2d)$ and $(15:26)$ which when removed by a 2-transformation give new tours A, B, C and D of lengths 168, 170, 171 and 178 respectively. However, solution C requires one of the vehicles to have a capacity of 22 tons and so is not feasible (figure 8.5b). A move to A, if chosen, allows a further move, to P, by the removal of the crossing $(13:26)$ (figures 8.5c, 8.6). Further exploration leads to the better solution Y (figures 8.5d, 8.6).

It should be clear how other feasibility checks on maximum distance, due date, delivery interval and other constraints may be incorporated. Also possible alternative transformations might be tried. Russell (1977) describes a routine, MTOUR, which uses the Lin and Kernighan (1973) approach to TSP with feasibility checks incorporated. It appears that MTOUR tends to give better quality results than most other heuristic methods that have been tried though, at least in some cases, it may require somewhat more computation time.

Savings Based Methods

In the B & B approach of example 8.1, a solution is obtained by building up routes from scratch, whereas Lin's approach starts with a feasible solution and small changes are made while maintaining feasibility. A third approach starts with each customer assigned to his own distinct route and then proceeds to combine routes, according to a 'savings' criterion, until the final solution is obtained. Two routes are only combined if doing so does not violate any capacity or maximum route length constraints that may be operative. The starting configuration is likely to be infeasible in that it requires more vehicles than are available; indeed it is not guaranteed that the final solution is feasible in this respect.

The total distance travelled in the symmetric one-customer-one-vehicle configuration is $2\sum_{j=1}^{n} d_{0j}$. If two customers, i and j say, are now assigned to one vehicle the route serving i and j has length $d_{0i} + d_{ij} + d_{j0}$. That is, there has been a saving of

$$S_{ij} = 2(d_{0i} + d_{0j}) - (d_{0i} + d_{ij} + d_{j0})$$
$$= d_{0i} + d_{0j} - d_{ij} \qquad (8.2)$$

It may be noted that two routes, each with one or more customers such that i is at one end of the first route and j is at one end of the second, combine to give the same saving S_{ij} (figure 8.7).

Clarke and Wright (1963) introduced the concept of 'savings' and a now classical procedure in the following algorithm.

Algorithm (CW)

Step 1 S_{ij} are calculated for all pairs of customers (i, j) and arranged in decreasing order.

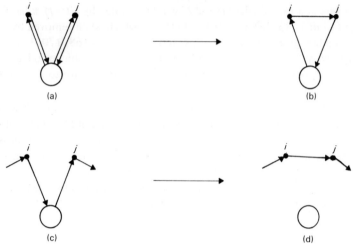

Figure 8.7

Step 2 Starting from the top of the list and working down then, two routes are combined by edge ij with saving S_{ij} if this is feasible with regard to the constraints of the problem; else, edge ij is rejected.

Example 8.3 Use Clarke and Wright's method to find a solution to the routing problem of example 8.1.

Solution The savings in decreasing size are

$S_{34} = 32, \quad S_{12} = 26, \quad S_{23} = 23, \quad S_{24} = S_{56} = 17$

$S_{13} = S_{16} = 13, \quad S_{14} = 8, \quad S_{26} = 6, \quad S_{15} = S_{36} = 4$

$S_{45} = 3, \quad S_{35} = S_{46} = 2, \quad S_{25} = 0$

Customers 3 and 4 are placed on the same route by adding edge 34 for a saving of 32 and a starting load of 7 tons. Next, customers 1 and 2 are placed on the same route by adding edge 12 for a saving of 26, and a starting load of 14 tons. Addition of edge 23 or 24 would join the first two routes and lead to the capacity of 15 tons being violated. Consequently edges 23 and 24 are rejected. The next edge to be added is 56 giving a saving of 17 and a new route with total load of 8.5 tons. No further edges can be added without violating the capacity constraint and so the final solution is

   depot–1–2–depot
   depot–3–4–depot
   depot–5–6–depot

which happens to be an optimal solution as was found in example 8.1.

The initial cost $2 \sum_{i=1}^{n} d_{0i}$ was 236 and the total savings $32 + 26 + 17 = 75$ giving a value of $236 - 75 = 161$ which can be checked to be correct by direct calculation of the combined lengths of the three routes.

Clarke and Wright's method tends to favour edges which are 'circumferential'. An alternative measure

$$\pi_{ij} = S_{ij} - d_{ij}$$

introduced by Gaskell (1967) has the effect of favouring more or less 'radial' edges which sometimes produces better quality routes (see figure 8.11). Yellow (1970) generalises to

$$\tilde{S}_{ij} = d_{0i} + d_{0j} - \gamma d_{ij}$$
$$= S_{ij} - (\gamma - 1) d_{ij}$$

and this form is adopted by Golden et al. (1977) who recommend solving a vehicle routing problem by the Clarke and Wright approach for several values of γ. This has the advantages that

(1) the quality of the best solution found will tend to be better than the 'traditional' approach which corresponds to taking $\gamma = 1$ only;
(2) a variety of 'sufficiently good' solutions allows the final selection to be made while taking into account 'unstated' goals and/or constraints (see chapter 11).

The basic Clarke and Wright approach also has the disadvantage that all savings S_{ij} are required. However, the addition of many edges is infeasible from a practical standpoint. Golden et al. (1977), to take this into account, first enclose the demand area by a rectangle with sides of length P and Q then divide this rectangle into K^2 sub-rectangles with sides P/K and Q/K. They restrict calculation of savings $\tilde{S}_{ij}$ to those pairs (i, j) for which j is in the same sub-rectangle as i or in one of the eight adjacent subrectangles (figure 8.8).

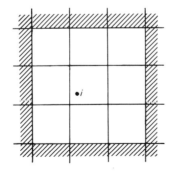

Figure 8.8

One of the shortcomings of the Clarke–Wright type of procedure is that an edge once added can never be removed. That is, the desirability of additions are treated as being independent of one another and account is not taken of the fact that the addition of one edge may preclude addition of certain other edges later on. This 'blocking' phenomenon arises in example 8.3 where edge 12 having been added all other edges $1j$, $2j$ must be rejected, and so it is not possible by these means to obtain any solution requiring only 2 vehicles. To remedy this a different approach was adopted by Holmes and Parker (1976). They first solve the problem by the basic Clarke–Wright approach (though refinements such as using the 'shape factor' γ and division of demand area into subrectangles can be incorporated readily). Next $i_1 j_1$, the first edge added to the incumbent solution, is temporarily suppressed and the problem re-solved. If a better solution is obtained then this becomes the new incumbent solution and $i_1 j_1$ is permanently suppressed; otherwise $i_1 j_1$ is added permanently. The next edge $i_2 j_2$ added to the incumbent solution is temporarily suppressed and so on until a prespecified number of suppressions results in no improvement.

Example 8.4 Apply the refinement process of Holmes and Parker to the routing problem of example 8.1, but with the customer requirements changed to $q_1 = q_2 = 7$, $q_3 = 3$, $q_4 = 5$, $q_5 = 2.5$ and $q_6 = 5$ tons. The maximum vehicle capacity remains at 15 tons and the distance matrix is unchanged.

Solution The savings are as in example 8.3. Algorithm (CW) adds edges 34, 12 and 56 as before and a total saving of 75 is achieved (figure 8.9). 34 is now

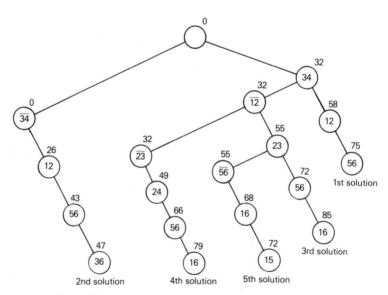

Figure 8.9 $\overline{ij}$ ($(\widehat{ij})$) indicate that edge ij is excluded (included)

temporarily suppressed and the problem re-solved, edges 12, 56 and 36 being added to get a total saving of 47. The first solution remains the incumbent and edge 34 is permanently added. Edge 12 is now temporarily suppressed and further edges 23, 56 and 16 added giving a total saving of 85.

At this point the third solution becomes the incumbent with edge 12 permanently suppressed. Edge 23 is now temporarily suppressed and the process continues with two further solutions being generated but with no further improvement. The third solution

   depot–2–3–4–depot
   depot–5–6–1–depot

with total length 236–85 = 151, is adopted.

Buxey (1979) notes that all heuristic methods for vehicle scheduling rely a great deal on time-consuming 'trial and error' evaluation procedures and that good solutions cannot be obtained without explicitly constructing many alternatives. For these reasons he argues that Monte Carlo simulation of feasible vehicle schedules might prove a powerful alternative! First a ranked table of savings is constructed excluding, if desired, savings for which the angle subtended at the depot exceeds some chosen angle θ_0 (for example, 90°). Next an edge ij is chosen with probability

$$p(ij) = \frac{(S_{ij})^M}{\Sigma^J (S_{ij})^M}$$

from the topmost J on the current savings list where M is some pre-set parameter and Σ^J indicates summation over the top J edges on the current savings list. Buxey noted that appropriate values of J and M are 5 and from 1 to 5 respectively. Conclusions are presented as the computer run time required for the best estimate to reach or surpass given values of total distance on either of 50 per cent or on 95 per cent of occasions. The numbers quoted do seem to substantiate Buxey's claim that Monte Carlo simulation provides a powerful method.

Mole (1979) noted a tendency of the savings approach to produce several routes which, when taken individually, fail to provide an adequate day's work for each vehicle but at the same time are sufficiently extensive to frustrate their combination. This suggests that sequential building of routes, to which we now turn, may have advantages.

8.2 ANGULAR APPROACHES TO VEHICLE ROUTING

As a consequence of theorem 8.2, solutions to vehicle routing problems, in the absence of capacity, maximum distance and due time constraints, will have a 'petal-like' appearance (figure 8.10a). However, capacity or maximum distance constraints can result in a certain amount of overlapping being profitable. For

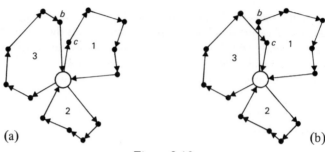

Figure 8.10

example, in figure 8.10a, suppose that the vehicle on route 3 is used to capacity and that customer b wishes to increase his order. In the new situation it is convenient for b to be reassigned to route 1 and for c to be reassigned to route 3 if feasible (figure 8.10b); although this leads to a slight increase in total route length, the new total route length may well be the minimum achievable.

Again, overlapping of routes may be desirable or even necessary if the number of vehicles is severely limited. A rather extreme example of this is provided by example 8.1 if only two vehicles are available.

For most cases an, at least approximate, petal-like solution is likely to be optimal. The vehicle routing problem can thus be reasonably tackled by dividing the demand area into sectors about the depot, such that each sector is serviced by a single vehicle. The route for each vehicle can then be found quite straightforwardly by an appropriate TSP solution procedure (Lin's method, for example). This approach is termed *sequential* or *cluster-first*. A notable contribution in this area is the SWEEP algorithm of Gillett and Miller (1974).

First the polar coordinates of the customer locations are computed taking the depot as origin and with polar angle relative to any fixed line from the depot. Some customer, k say, called the *seed*, is selected. A ray from the depot to customer k (conceptually) sweeps through 360° clockwise (or anticlockwise). Starting with k, customers are assigned as they are swept first to vehicle 1 as long as capacity constraints permit, then to vehicle 2 and so on. The route of each vehicle can be specified by taking customers in the order in which they were assigned and using Lin's λ-opt method. Finally, the routes may be capable of further refinement by allowing interchanges of customers between routes.

The process can be repeated for a variety of choices of seed; in the extreme case every customer is chosen as seed for both clockwise and anticlockwise sweeping.

Example 8.5   Apply the SWEEP algorithm to the routing problem of

(1) example 8.1, and
(2) example 8.4.

Distribution Problems

Solution (1) We consider clockwise sweeping only and list the routes obtained in the first phase for all choices of seed. (D denotes 'depot'.)

Seed	Routes			Total length
1	D–1–2–D	D–3–4–5–D	D–6–D	175
2	D–2–3–D	D–4–5–6–D	D–1–D	193
3	D–3–4–5–D	D 6 1 D	D 2–D	208
4	D–4–5–6–D	D–1–2–D	D–3–D	190
5	D–5–6–1–D	D–2–3–D	D–4–D	183
6	D–6–1–D	D–2–3–D	D–4–5–D	197

The optimal 3-vehicle solution is not in this list but is obtained in the next phase from the first solution by moving customer 5 to the third route.
(2) This data set shows the SWEEP algorithm to better advantage with the optimal 2-vehicle solution being obtained when the seed is 2 or 5 for clockwise sweeping (1 or 4 for anticlockwise sweeping).

Mole and Jameson (1976) noted that building routes sequentially (rather than concurrently as in Clarke and Wright's method, for example) permits the freedom of scheduling vehicles as routes are planned, thus allowing possibilities for several out and return journeys by each vehicle to be actively pursued. They used the generalised saving

$$S_{ij}(x) = 2d_{0x} + d_{ij} - d_{ix} - d_{jx}$$

of inserting customer x, who at this stage is on a route by himself, into the emerging route between successive vertices i and j. The more general $S_{ij}(x)$ were employed by repeating the following three steps.

Step 1 For each customer x, who is still on a route of his own, the most advantageous point of insertion on the emerging route is determined by finding

$$S(x) = \min S_{ij}(x).$$

Step 2 A customer x, whose saving $S(x)$ is maximal, is placed on the emerging route in the position indicated in step 1.

Step 3 The emerging route is transformed so as to maintain 2-optimality.

Notice that the improvement transformations are carried out as routes are being built and that savings $S_{ij}(x)$ need not be updated once found, though values of $S_{ij}(x)$ for new values of i and j will be required as the algorithm progresses. Published results suggest that this method leads, on average, to lower quality results than SWEEP algorithms. However, the method is very quick and it would appear that incorporation of the speeding up techniques of Golden *et al.* (1977) (see section 8.1) would lead to a method viable for very large problems while giving, on average, better solutions than obtained using the basic (concurrent) form of the Clarke–Wright method.

Foster and Ryan (1976) took the idea of petal-shaped routes a step further and developed a 0–1 ILP formulation which determines the optimal set of non-overlapping petal-shaped routes (each serving its own sector). They then relaxed this by allowing a certain amount of overlapping. The results obtained were of very high quality being comparable to those of Russell (1977) and being obtained for a comparable expenditure of computational effort.

At the time of writing no exact method is known which can solve other than very small problems (no more than 25 to 30 customers). For problems of up to 100 or so customers it appears that MTOUR and Foster and Ryan's method provide the best quality results within a few seconds computing time. For medium sized problems (up to around 250 customers) the modified SWEEP algorithm will provide a good solution in a manageable computing time. For very large problems the method of Golden *et al.* (1977), a streamlined version of Mole and Jameson's method, or Monte Carlo simulation may be used to obtain solutions.

We conclude this section with an example of a vehicle scheduling problem in which the vehicles are boats and travel is in straight lines between customers.

Example 8.6 Oil wells in a large lake have to be sampled by withdrawing, near the well head, a fraction of the liquid flowing from the well. The samples are then taken to a laboratory to determine the important oil/liquid ratio. The sampling is not performed at fixed intervals, but with a frequency that varies from well to well and also changes with time.

A list is provided of the N wells to be sampled in the following week. Each well has an associated priority and the list is arranged in order of decreasing priority. (Since priority will increase with overdue time every well will be sampled in due course.)

Given a fleet of L launches find a set of feasible routes that minimises the total cost of visiting all of the wells on the list, subject to an 8 hour limit on the time taken on any of the routes.

Solution This problem is a simplification of the one described by Cunto (1978) which related to the 3000 oil wells of Lagoven, S.A. in Lake Maracaibo, Venezuela. The lake was divided into two sections each normally served by one launch 5 days a week. That is, the problem is decomposed into two smaller ones with N up to 400 and $L = 5$. The approach given below is an outline version of that given by Cunto and for further details of the problem and solution method readers are referred to his paper.

First, n top priority wells are considered, where n is a conservative estimate of the number that can be sampled.

Next the wells are divided into groups and a travelling salesman routine used

Distribution Problems

to provide a route for each group. If the slack time permits further samples to be taken then n is increased (if less than N) and a new set of routes produced. If n reaches N then a full voyage is deactivated if possible and the assigned wells reallocated.

When n cannot be increased further, or if $n = N$ and no more full voyages can be deactivated, a travelling salesman routine is applied to all routes in the current solutions.

Initial solutions are obtained, for the chosen n wells, using two approaches

(1) an 'angular' approach following Gillett and Miller;
(2) a clustering approach.

The whole problem is solved using each of these approaches for generating initial solutions and the better of the two resulting solutions is then the one finally adopted.

The merits of the angular approach have been dealt with earlier. However, if points are distributed in clusters with more than one cluster in a given direction, call sets derived by angular means can lead to poor initial solutions. Figure 8.11 provides such an example. Cunto encountered this clustering effect in one of the two sections of Lake Maracaibo and devised the following clustering algorithm.

Step 1 Find the well x furthest from the origin which is as yet not assigned to a voyage.

Step 2 Grow an MST from x [as in algorithm (P)] with each well added to the MST being assigned to the same voyage.

Step 3 When a voyage has been made up the process stops and is repeated for the next voyage and so on until all voyages are complete.

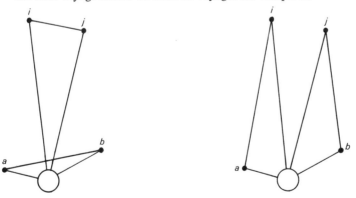

Figure 8.11 The routes on the left, obtained by clustering, have lower total length than the ones on the right

Two improvement heuristics (see chapter 11) were employed in Cunto's method and of the four described two were computationally effective. In the first, single points were transferred from one route to another route 'close to that point' if a reduction in distance resulted. In the second, two routes were replaced by two new routes, which took account of clustering better, if a reduction in distance resulted.

Cunto's method appears to be very effective as judged by the quality of solution reported for a well-known test problem and by the computational effort required.

For the well sampling application an interesting modification was introduced. It was noted that

(1) the travelling salesman routine required a considerable proportion of the time to obtain the initial sequence of points within a route;
(2) a route that initially violates the time constraint could, as a result of the improvement process, become feasible at a later stage.

For these reasons, it was decided to relax the time constraint by allowing launches to operate overtime, a common practice in field operations. Correspondingly the cost function was modified to charge overtime at a premium rate. Also, if the current solution demands overtime no further increase in n is permitted nor are voyages deactivated.

The system, incorporating the method briefly described, was installed in April 1975 and has been in weekly operation since then. Advantages of the system are

(1) better timing of samples with undersampling and oversampling considerably reduced;
(2) a net saving of about 10 hours per week supervisory personnel time which can be devoted to more critical production costs;
(3) better instructions for the operators;
(4) fewer voyages are required, with one full voyage out of every five being eliminated.

8.3 MULTI-DEPOT DISTRIBUTION PROBLEMS

Though many papers have been written on single-depot distribution problems the multi-depot generalisation of VRP has received relatively little attention. Exact methods for VRP can be extended to the general case, but in view of the computational difficulties encountered with the single-depot problem it will be appreciated that this is not practicable for other than the smallest multi-depot problems.

Golden *et al.* (1977) distinguish three classes of heuristic method that have been applied to multi-depot problems: improvement heuristic methods;

generalisation of the SWEEP algorithm; generalisation of the Clarke–Wright savings approach.

(a) Improvement Heuristic Methods

Wren and Holliday (1972) noted the problems of extending single-depot methods if an individual customer may be served by one of several depots. Correspondingly the restriction was made that each customer be assigned to a unique depot. Once this assignment has been made the routes for individual depots may then be made by the methods of sections 8.1 and 2.

The initial assignment of customers to depots is a non-trivial task. It could be treated as a standard transportation problem, but this is unrealistic as the costs are not proportional to the quantity of goods delivered to a customer and the most appropriate depot to which to assign a customer will depend on the relative positions of neighbouring customers (exercise 8.6). If all depots are considered together a better solution is likely to result than if the problem is decomposed into a number of single-depot VRPs.

Wren and Holliday (1972) first *provisionally* assigned each customer to the nearest depot and for each depot sorted customers in order of polar angle relative to a line from that depot in a direction in which the customers are most sparse. Four different starting solutions are obtained starting from polar angles 0°, 90°, 180° and 270°. From these, four improved solutions are obtained and the best of these is finally selected as the solution to the problem.

Improvements are made in a variety of ways by: 2-opt transformations; moving a customer to another point on the same route or another route (possibly from another depot); moving pairs of customers; deletion of a route and redistributing the customers it served; untangling 'tangled' routes.

When tested on well-known single-depot test problems Wren and Holliday's method led to good quality results for the expenditure of a modest amount of effort. However, in the small to medium range it has probably been superseded by the SWEEP algorithm for the single depot VRP.

For the multi-depot case Cassidy and Bennett (1972), using a similar approach, reported a successful application to the delivery of school meals.

(b) Modified SWEEP Algorithm

This method, due to Gillett and Johnson, solves the multi-depot problem in two stages. Initially all customers are in an 'unassigned' state. Then for each customer i the ratio

$$r(i) = \frac{d_{i\alpha'}}{d_{i\alpha''}}$$

is computed, where α' and α'' are the closest and second closest depots to i

respectively. Customers are assigned to the nearest depot in increasing order of $r(i)$ until ratios close to 1 are reached. The remaining customers are treated more carefully, a customer being assigned to a depot (by being inserted on an emerging route) in the least costly manner. A number of refinements are made to improve the resulting solution.

(c) Savings Based Approaches

Tillman and Cain (1972) used a modified Clarke–Wright method in which a customer is temporarily first assigned to the nearest depot; the assignment only becomes permanent when the customer is on a route containing two or more customers. The saving S^k_{ij} associated with customers i and j and depot k is

$$S^k_{ij} = \tilde{d}^k_i + \tilde{d}^k_j - d_{ij}$$

where

$$\tilde{d}^k_i = \begin{cases} 2 \min_t (d_{it}) - d_{ik} & \text{if } i \text{ has a temporary label (minimisation is over depots } t \text{ only)} \\ d_{ik} & \text{otherwise} \end{cases}$$

The first form for $\tilde{d}^k_i$ corresponds to a customer being reassigned from the closest depot to the same or another depot. The method is time-consuming since the S^k_{ij} are computed for all k and all distinct pairs i and j.

Tillman and Cain also define a penalty p^k_{ij} for not linking i and j on a route from k, and customers are assigned to routes, and hence depots, in decreasing order of the generalised saving

$$\bar{S}^k_{ij} = \alpha S^k_{ij} + \beta p^k_{ij}$$

where α and β are two appropriately selected positive weights.

As noted by Golden et al. (1977) penalties could also be used for the single-depot VRP thus improving on Clarke and Wright's original method.

Golden et al. adopted the Tillman–Cain approach using a superimposed grid with savings S^k_{ij} considered only if i and j are sufficiently close (section 8.1). This together with some programming refinements constituted their algorithm I. This method was then modified so as to be applicable to large problems. Customers for whom the Gillett–Johnson parameter $r(i)$ (see above) is small, that is, $0 \leq r(i) < \delta \leq 1$ for some preassigned value of the parameter δ, are assigned to their nearest depot. The remaining customers are then assigned to depots as in algorithm I. Finally the solution is produced, depot by depot, using single depot VRP techniques.

Solutions can be refined by transferring a single vertex at a time from one route to another.

Golden et al. were able to obtain results comparable in quality to ones obtained for the same problems using the Gillett–Johnson approach outlined above. However, considerably less computation was required.

Distribution Problems

The models discussed above are adequate for some problems but fail to include many features of interest in real distribution problems. Related aspects include

(1) due dates for deliveries;
(2) variable customer demand;
(3) vehicle loading problems;
(4) fleet size determination taking account of vehicle overhead costs and cost of hiring;
(5) allowing routes to start from one depot and finish at another;
(6) inter-relationship between location and distribution problems.

For a further discussion of some of these aspects of distribution problems readers are referred to Eilon et al. (1971).

EXERCISES

8.1 In the solution to parts (1) and (3) of example 8.1, several arcs were excluded on the grounds that pseudo depots are indistinguishable. Verify that this cannot eliminate all optimal solutions (whatever the values of c_{ij}). Extend to the cases of $5, 6, \ldots, m$ pseudo depots.

8.2 Given the distance matrix and customer requirements vector of example 8.1, find routes of minimal total length if there are two vehicles with capacities of
(a) 15 and 20 tons; and
(b) 15 and 22 tons.

8.3 Four customers 1, 2, 3, 4 with requirements of $q_1 = 2, q_2 = 5, q_3 = 5, q_4 = 2$ units are to be supplied from a central depot by two vehicles each with capacity $Q = 10$ units. The route for a single vehicle is not to exceed 20 miles, and no load-splitting is allowed. Use a savings approach to find a solution to this problem for the network in figure 8.12 in which the number alongside edge ij gives the length in miles of the corresponding journey.

By developing a suitable enumeration tree, or otherwise, show that the problem has no feasible solution if q_2 is changed to 6 units (the remaining data being unchanged). (Note that it may be found useful to demonstrate first that one vehicle must be assigned to customer 3 alone.)

(LU 1978, abridged)

8.4 M depots are to be established in a given district to serve n customers. Customer i has a requirement q_i and each vehicle a capacity Q. The problem is to site the M depots, each allocated as many vehicles as may be needed, at the vertices or on the edges of the network connecting them so that the total travel distance is minimised. Prove that there is an optimal solution in which all depots are located at vertices.

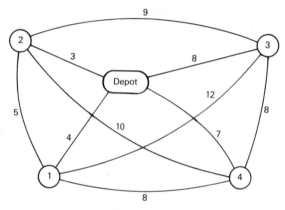

Figure 8.12

8.5 Devise an algorithm which will solve the location problem of exercise 8.4.

8.6 In a single-commodity multi-depot situation it is not always optimal, with regard to overall distance travelled, to assign customers to their nearest depots. Consider the two-depot problem with depots at coordinate points (0, 0) and (2, 2) and customers at (0, 1), (0, 3), (1, 4), (3, 4), (4, 3), (4, 1), (3, 0) and (1, 0). Verify that, in the absence of capacity constraints, it is optimal to assign all customers to the first depot. Keeping the first depot at (0, 0) find the set of points (x, y) at which the other depot might be located so that there is an optimal solution in which both depots have customers assigned to them.

8.7 A large company has operations at several sites, and frequently requires to transport goods between the sites. That is, each site acts as a 'depot' and a 'customer' in a multi-commodity set up. By way of example, consider the case of three sites distance D from each other with an amount t_{ij} needing to be transported from site i to site j where

$$(t_{ij}) = \begin{bmatrix} 0 & 14 & 0 \\ 7 & 0 & 6 \\ 7 & 3 & 0 \end{bmatrix}$$

Vehicles have a capacity of 10 units. If vehicles make only direct trips between two sites then four journeys are required (two between sites 1 and 2) leading to a total distance travelled of $8D$. However, if vehicles can visit more than two sites on a journey but goods only travel between successive sites then the total distance travelled can be reduced to $7D$ by assigning goods to routes as follows

$$1 \xrightarrow{10} 2 \xrightarrow{0} 3 \xrightarrow{7} 1 \quad 2 \xrightarrow{7} 1 \xrightarrow{4} 2 \quad 3 \xrightarrow{3} 2 \xrightarrow{6} 3$$

the numbers above the arcs giving the quantity of goods carried on that section.

If goods are allowed to be 'carried through' sites then the total distance travelled can be further reduced to $5D$

$$2 \xrightarrow{6} 3 \xrightarrow{7} 1 \xrightarrow{4} 2 \qquad \overset{3}{\frown} \qquad 1 \xrightarrow{10} 2 \xrightarrow{7} 1$$

Consider now the problem involving four sites with inter-site distances in miles given by matrix (d_{ij}) and transportation requirements given by (t_{ij})

$$(d_{ij}) = \begin{bmatrix} 0 & 65 & 104 & 53 \\ & 0 & 77 & 101 \\ & & 0 & 98 \\ & & & 0 \end{bmatrix} \quad (t_{ij}) = \begin{bmatrix} 0 & 30 & 3 & 11 \\ 4 & 0 & 14 & 24 \\ 12 & 7 & 0 & 6 \\ 3 & 17 & 0 & 0 \end{bmatrix}$$

If there are two vehicles with capacities 20 units based at each of the sites and no journeys may exceed 250 miles in length, list the possible routes that vehicles may take. Verify that the overall travel distance is at least 1328 miles if only journeys between two sites are allowed. Find routes and assignments of goods to routes so that this overall distance is reduced to 1100 miles or less if journeys involving more than two sites are permitted but carry through is not. Show that if carry through is also permitted then the overall distance may be reduced to under 900 miles.

Suppose that such routing problems arise daily. Devise a heuristic algorithm which will lead to good route assignments.

9 Flows in Networks: Basic Model

Many problems in OR are concerned with the flow of some 'substance' or 'set of items' through a network. For example, the flow of vehicles through a road system, trains through a railway network, messages through a communication network, fluids through pipes, electricity through an electrical circuit, people through customs at a large airport, jobs capable of being produced on different machines through their production sequence. There is usually an upper limit, and sometimes a non-zero lower limit as well, on the magnitude of the flow along each arc of the network. This in turn leads to restrictions on the total flow through the network and it is this aspect which is introduced in section 9.1. A basic model is put forward with various extensions being dealt with in chapter 10. The max-flow min-cut theorem and some of its consequences are given in section 9.1, and in section 9.2 the Ford–Fulkerson algorithm for obtaining a maximal static flow is presented together with some refinements due to Dinic (1970), Karzanov (1974) and Pohl (1977).

9.1 COMPLETE FLOWS AND MAXIMAL FLOWS

Consider the case of a section of dual carriageway road which can support, in good conditions (light vehicles, good weather, etc.), a steady stream of traffic in either direction of up to (say) 2500 vehicles per hour above which it 'chokes up'. Then the flows f_1, f_2 along each carriageway can be as little as 0 vehicles per hour and in good conditions as much as $c = 2500$ vehicles per hour; that is, $0 \leqslant f_1 \leqslant c, 0 \leqslant f_2 \leqslant c$.

Suppose now that a network of roads is given with c specified for each section of road, what is the maximum achievable net flow of traffic from one point S to another point F through the network? In order to deal with this type of problem and to construct an appropriate model it is convenient first to give some definitions.

Definition 9.1
A network, in which each arc ij has an associated *capacity* $c_{ij} > 0$, with a unique source S and a unique sink F will be termed *standard*.

Definition 9.2
An *(arc) flow* in arc ij is a real number f_{ij} satisfying $0 \leqslant f_{ij} \leqslant c_{ij}$. A *(network*

Flows in Networks: Basic Model

flow from S to F is a set of arc flows $\{f_{ij}\}$, and is *conserved* if for each vertex j, other than S and F, the flow into j balances the flow out of j; that is

$$\sum_{i \in \Gamma^{-1}j} f_{ij} = \sum_{k \in \Gamma j} f_{jk} \qquad \text{all } j \neq S, F$$

Note that when 'flow' is used without a qualifying adjective the precise meaning intended should be clear from the context.

Definition 9.3
The *value* ϑ of a network flow $f = \{f_{ij}\}$ is the total flow out of S (or into F since flow is conserved), that is

$$\vartheta = \sum_{a \in \Gamma S} f_{Sa} = \sum_{b \in \Gamma^{-1} F} f_{bF}$$

We are now in a position to present the basic problem of network flow theory.

Basic Problem Find a maximal valued conserved flow in a directed standard network.

Written out in full the basic problem is seen to be the following linear programming problem

$$\text{maximise} \quad \vartheta = \sum_{a \in \Gamma S} f_{Sa} \tag{9.1}$$

subject to

conservation $\quad \sum_{i \in \Gamma^{-1}a} f_{ia} - \sum_{k \in \Gamma a} f_{ak} = 0, \qquad a \neq S, F \tag{9.2a}$

capacity $\quad\quad\quad f_{ij} \leq c_{ij} \quad\quad \text{all arcs } ij \tag{9.2b}$

non-negativity $\quad f_{ij} \geq 0 \quad\quad \text{all arcs } ij \tag{9.2c}$

As indicated, the constraints 9.2 will be known as the *conservation*, *capacity* and *non-negativity* constraints respectively.

Example 9.1 (A bus scheduling problem from Saha, 1970) Each division of an Indian State Transport Corporation operates a fleet of, on average, 300 buses, for which timetables have to be prepared three times a year. Having assessed the demand for buses, a list of trips of the form $\alpha = (P_S^\alpha, P_F^\alpha, T_S^\alpha, T_F^\alpha)$ is determined where for trip α, P_S^α and P_F^α are the start and destination points, and T_S^α and T_F^α are the start and finish times. A single bus can service trips α and β consecutively provided $P_F^\alpha = P_S^\beta$ and $T_F^\alpha \leq T_S^\beta$ thereby leading to the grouping of trips. Ideally this grouping should be such that the number of buses is minimised. Saha states that the equivalent of two man-weeks was spent on grouping trips at each scheduling, and this led him to look for a suitable algorithm for mechanising the procedure. In his investigation he produced the following network flow formulation.

Solution Define $H = (X, A, c)$ to be the directed network with

$$X = \{S, a_\alpha, b_\beta, F\}, \qquad \alpha, \beta = 1, \ldots, n$$
$$A = \{Sa_\alpha, b_\beta F \mid \alpha, \beta = 1, \ldots, n\} \cup \{a_\alpha b_\beta \mid \beta \text{ can follow } \alpha\}$$

where n is the number of trips. Note that it may be found convenient to include 'dummy' trips from depots at the start of the day, and to depots at the end of the day. All arcs are assigned a unit capacity. A flow of 1 in $a_\alpha b_\beta$ is understood to mean that a single bus can service trip β immediately after trip α. Constraints on flows in arcs Sa_α and $b_\beta F$ relate to the restriction that one bus can service only one trip at a time and any trip is serviced by only one bus. Specifically, the constraints are

$$\sum_\beta f_{a_\alpha b_\beta} = \text{flow into } a_\alpha = f_{Sa_\alpha} \leqslant 1 \tag{9.3a}$$

$$\sum_\alpha f_{a_\alpha b_\beta} = \text{flow out of } b_\beta = f_{b_\beta F} \leqslant 1 \tag{9.3b}$$

$$f_{Sa_\alpha}, f_{b_\beta F}, f_{a_\alpha b_\beta} \leqslant 1 \qquad \text{all } \alpha, \beta \leqslant n \tag{9.4}$$
$$f_{Sa_\alpha}, f_{b_\beta F}, f_{a_\alpha b_\beta} \geqslant 0 \qquad \text{all } \alpha, \beta \leqslant n \tag{9.5}$$

Now the number of buses required is equal to $n - \sum_{\alpha,\beta} f_{a_\alpha b_\beta}$ and so

$$\vartheta = \sum_{\alpha,\beta} f_{a_\alpha b_\beta} = \sum_\alpha f_{Sa_\alpha}$$

must be maximised subject to conservation constraints 9.3, the capacity constraints 9.4 and the non-negativity constraints 9.5 That is, this bus scheduling problem has been formulated as a network flow problem of the basic form given by 9.1 and 9.2. It may be noted that the specially simple structure of the bus scheduling problem admits a more efficient solution than can be obtained by directly using the Ford–Fulkerson algorithm (section 9.2) and the interested reader is referred to Saha's paper.

In general a network with n vertices and q arcs leads, as above, to a linear programming formulation with q variables f_{ij} and $n-2+2q$ constraints (including the q non-negativity constraints). Thus, even a moderate sized network can lead to a fairly large linear programming problem, and it would seem preferable to adopt an alternative approach.

Example 9.2 Find a maximal flow from S to F for the network shown in figure 9.1.

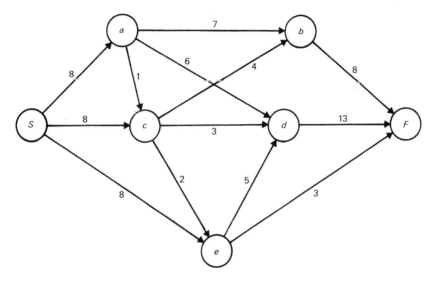

Figure 9.1 The capacity c_{ij} is given alongside each arc ij

Solution We shall first try to solve this problem using an approach which will be termed the *intuitive method*. The starting point is some geometrical representation of the network (in this case figure 9.1). The elementary paths from S to F are then considered in order of 'superiority', assigning in turn the largest flow possible to each.

Even for this small problem it is desirable to adopt a systematic treatment and a scheme suitable for manual solution is aided by keeping a table such as shown in table 9.1. The *residual* capacity of an arc is just its original capacity minus the flow assigned to it.

The (most) *superior* path is obtained by starting at S and at each vertex choosing the leftmost outgoing arc until F is reached. (If a dead end is encountered or a vertex is encountered for the second time, then the path is traced back to the first vertex which has an untried outgoing arc and the path from that chosen). For the example network the superior path is $SabF$, and it is seen that the flow in it will be limited by the capacity $c_{ab} = 7$, that is ab is a bottleneck arc for this path. Accordingly a flow of 7 is sent along $SabF$ by increasing f_{Sa}, f_{ab} and f_{bF} each by 7 units and adjusting the residual capacities (table 9.1). Arc ab now has zero residual capacity and is said to be *saturated*.

At each stage the next most superior path from S to F is the superior path from S to F in the network with all saturated arcs disregarded. For the example network $SadF$ is the second most superior path. Sa is a bottleneck arc with residual capacity 1 and so 1 unit of flow is assigned to $SadF$. Next 1 unit of flow

Table 9.1

Arc	Capacity	Residual capacity	Arc flow
Sa	8	~~8~~ ~~1~~ 0	~~0~~ ~~7~~ 8
Sc	8	~~8~~ ~~7~~ ~~4~~ 2	~~0~~ ~~1~~ ~~4~~ 6
Se	8	~~8~~ ~~3~~ 2	~~0~~ ~~3~~ 6
ab	7	~~7~~ 0	~~0~~ 7
ac	1	1	0
ad	6	~~6~~ 5	~~0~~ 1
cb	4	~~4~~ 3	~~0~~ 1
cd	3	~~3~~ 0	~~0~~ 3
ce	2	~~2~~ 0	~~0~~ 2
bF	8	~~8~~ ~~1~~ 0	~~0~~ ~~7~~ 8
ed	5	~~5~~ ~~3~~ 0	~~0~~ ~~2~~ 5
eF	3	~~3~~ 0	~~0~~ 3
dF	13	~~13~~ ~~12~~ ~~9~~ ~~7~~ 4	~~0~~ ~~1~~ ~~4~~ ~~6~~ 9

Path	Flow assigned
$S \to a \to b \to F$	7
$S \to a \to d \to F$	1
$S \to c \to b \to F$	1
$S \to c \to d \to F$	3
$S \to c \to e \to d \to F$	2
$S \to e \to d \to F$	3
$S \to e \to F$	3
Total flow assigned	20

is assigned to *ScbF* and so on as indicated in table 9.1 until the flow pattern shown in figure 9.2 is obtained, at which point it is seen that no further (strictly positive) path flows can be assigned. Such a flow is called *complete*.

Although complete the given flow is not maximal and so the intuitive method has failed to solve the given problem. To see that this flow of value 20 is non-maximal note that the removal of 2 units of flow from *SabF* and the addition of 2 extra units of flow to each of *SadF* and *ScbF* leads to a feasible flow of value 22.

The Max-flow Min-cut Theorem

In order to describe a practical algorithm for augmenting non-maximal flows it is convenient to show the prominent role that cuts [partitions of the vertex set which separate source and sink (see section 6.3)] play in the theory of network flow.

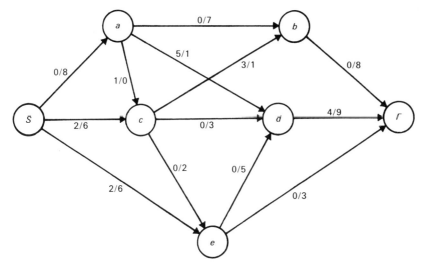

Figure 9.2 The numbers alongside each arc ij are of the form r_{ij}/f_{ij} where r_{ij} is the residual capacity of ij

Definition 9.4
The *capacity* $c(A,\bar{A})$ of a cut $(A,\bar{A})$ in a network is

$$c(A,\bar{A}) = \sum_{\substack{i \in A \\ j \in \bar{A}}} c_{ij}$$

Theorem 9.1
The value ϑ of a network flow in a standard network is bounded by the capacity of any cut and so

$$\vartheta \leq \min_{(A,\bar{A})} c(A,\bar{A})$$

Proof It is evident that any flow of substance from source to sink must all pass through arcs ij with $i \in A$ and $j \in \bar{A}$. Since flow in arcs from $\bar{A}$ to A cannot increase the value of the flow the result follows. □

Definition 9.5
Given a flow pattern $f = \{f_{ij}\}$, then a special partition $(A_f, \bar{A}_f)$, the partition *associated* with f, is defined recursively by

(1) place S in A_f; and
(2) for each j already in A_f place k in A_f if $f_{jk} < c_{jk}$ and place i in A_f if $f_{ij} > 0$,

and setting $\bar{A}_f = X - A_f$ where X is the set of all network vertices.

Example 9.3 Determine the partition associated with the complete flow illustrated in figure 9.2.

Solution Initially set $A_f = \{S\}$, and adjoin c and e since Sc and Se are not saturated. Then b is adjoined since cb is unsaturated and a is adjoined since ab is not *flowless*. As ad is unsaturated, d is adjoined, and finally F is adjoined as dF is unsaturated.

$$(A_f, \bar{A}_f) = (\{S, a, b, c, d, e, F\}, \phi)$$

Theorem 9.2
If f is a maximal flow of value ϑ, then $(A_f, \bar{A}_f)$ is a cut.

Proof As, by definition, $S \in A_f$ it is only necessary to show that maximality of f implies $F \in \bar{A}_f$.

Suppose that F is not in $\bar{A}_f$ then it must be in A_f. From the constructive definition of A_f it can be seen that there must be a chain $\pi \equiv u_0 u_1 \ldots u_p$ from $S = u_0$ to $F = u_p$ which lies entirely in A_f and such that $u_{\alpha+1}$ is added to A_f by considering arc $u_\alpha u_{\alpha+1}$ or $u_{\alpha+1} u_\alpha$. An arc $u_\alpha u_{\alpha+1}$ of π, whose direction accords with traversing π from source to sink will be termed a *forward* arc, and an arc $u_{\alpha+1} u_\alpha$ of π will be termed a *backward* arc. Let $FA(\pi)$ and $BA(\pi)$ denote the set of forward and backward arcs on π, and

$$\epsilon_\alpha = \begin{cases} \text{residual capacity } r_{u_\alpha u_{\alpha+1}} & \text{if } u_\alpha u_{\alpha+1} \in FA(\pi) \\ \text{flow } f_{u_{\alpha+1} u_\alpha} & \text{if } u_{\alpha+1} u_\alpha \in BA(\pi) \end{cases}$$

From the way in which the chain π is constructed $\epsilon_\alpha > 0$, $\alpha = 0, 1, \ldots, p-1$ and hence $\epsilon = \min_\alpha \epsilon_\alpha$ is also strictly positive. The value of the flow can now be increased by augmenting the flow along π

$$\begin{cases} f_{u_\alpha u_{\alpha+1}} \text{ is increased by } \epsilon & \text{if } u_\alpha u_{\alpha+1} \in FA(\pi) \\ f_{u_{\alpha+1} u_\alpha} \text{ is decreased by } \epsilon & \text{if } u_{\alpha+1} u_\alpha \in BA(\pi) \end{cases}$$

Since this change violates none of the constraints 9.2 the new flow pattern is feasible and its value is strictly greater (by ϵ) than ϑ. But this contradicts the maximality of f and so the assumption that F is in A_f must be false. The desired result follows. □

A chain from S to F which 'lies entirely in A_f' will be termed *flow augmenting*. If f is maximal then $(A_f, \bar{A}_f)$ is the *associated* cut of f.

Example 9.4 From example 9.3 it is seen that $ScbadF$ is a flow augmenting chain. The variables ϵ_α of theorem 9.2 are $\epsilon_0 = 2$, $\epsilon_1 = 3$, $\epsilon_2 = 7$ (backward arc), $\epsilon_3 = 5$, $\epsilon_4 = 4$ and $\epsilon = 2$. The flows in Sc, cb, ad and dF are increased by 2 units. The new flow pattern shown in figure 9.3a has value 22.

Flows in Networks: Basic Model

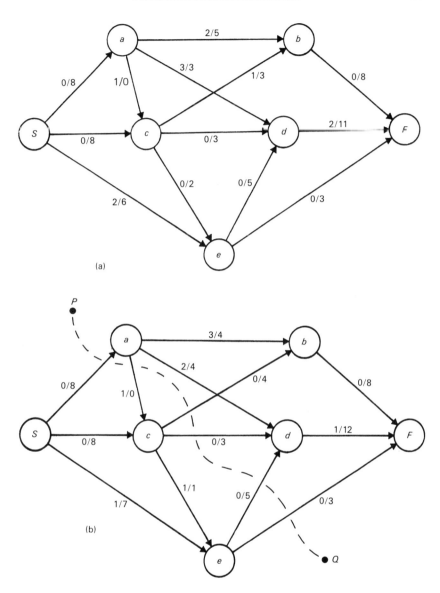

Figure 9.3

Again $A_f = \{ S, a, b, c, d, e, F \}$ and so the flow is not maximal. *SecbadF* is a flow augmenting chain and

$\epsilon = \min (2,2,1,5,3,2) = 1$

The flow is augmented by 1 unit (figure 9.3b).

Calculating the associated partition for the new flow yields

$$(A_f, \bar{A}_f) = (\{S, c, e\}, \{a, b, d, F\})$$

Thus $F \in \bar{A}_f$, $(A_f, \bar{A}_f)$ is a cut, and it will now be shown that this implies that f is maximal.

Theorem 9.3 (Max-flow Min-cut)
If ϑ_f denotes the value of flow f in a standard network then

$$\max_f \vartheta_f = \min_{(A,\bar{A})} c(A, \bar{A})$$

Proof By theorem 9.2, it is only necessary to find a cut $(B, \bar{B})$ for which $\max \vartheta_f = c(B, \bar{B})$.

Let $f = \{f_{ij}\}$ be some maximal flow and $(A_f, \bar{A}_f)$ the associated cut. Any arc ij from A_f to $\bar{A}_f$ must be saturated otherwise j would be in A_f. Similarly any arc kl from $\bar{A}_f$ to A_f must be flowless. Hence

$$\vartheta_f = \sum_{\substack{i \in A_f \\ j \in \bar{A}_f}} f_{ij} - \sum_{\substack{k \in \bar{A}_f \\ l \in A_f}} f_{kl} = \sum_{\substack{i \in A_f \\ j \in \bar{A}_f}} c_{ij} - 0 = c(A_f, \bar{A}_f)$$

The result follows immediately. □

Corollary 1 if $(B, \bar{B})$ is any minimal cut and f a maximal flow then every arc from B to $\bar{B}$ is saturated and every arc from $\bar{B}$ to B is flowless.

Corollary 2 Under the conditions of corollary 1, $A_f \subset B$.

Example 9.5 For the network of example 9.1, reproduced in figure 9.3b, a minimal cut is $(\{S, c, e\}, \{a, b, d, F\})$ and is illustrated by the broken separating line from P to Q. The capacity of the cut is $8 + 4 + 3 + 5 + 3 = 23$ which is just the value of a maximal flow as required.

It has been seen that the intuitive method can lead to complete flows which are not maximal. However, we are now in a position to show that the intuitive method can be applied to a planar graph in such a way that a maximal flow is assured. In order to do this the concept of 'irreducibility' is introduced.

A cut $(D, \bar{D})$ is a *reduction* of cut $(B, \bar{B})$ if the set of arcs from D to $\bar{D}$ is a subset of the set of arcs from B to $\bar{B}$. A cut is *irreducible* if it has no reductions other than itself.

Lemma Let N be a standard network geometrically represented without crossings. If $(A, \bar{A})$ is an irreducible cut then the superior path from S to F includes just one arc from A to $\bar{A}$.

Flows in Networks: Basic Model

Proof (The proof relies heavily on geometric visualisation and readers are advised to construct example networks to satisfy themselves that the arguments are valid.)

Let $(B, \bar{B})$ be any cut not necessarily irreducible. Take two points P and Q on the unbounded face for the given representation. A line is constructed from P to Q, to separate B and $\bar{B}$, such that the only arcs of N which are intersected are those between B and $\bar{B}$. (An example is afforded by the broken line PQ of figure 9.3b for the network obtained by removal of arc ad.) We may also assume (by interchanging P and Q if necessary) that the first elementary path from S to F that PQ intersects is the superior path.

Suppose PQ intersects the superior path first at K on an arc from B to $\bar{B}$ then at L on an arc from $\bar{B}$ to B then at M on an arc from B to $\bar{B}$. A new cut $(D, \bar{D})$ can be defined by joining P to M without intersecting the superior path then continuing from M to Q as before (figure 9.4). It can be seen that $(D, \bar{D})$ is a reduction of $(B, \bar{B})$. It follows that for any cut $(A, \bar{A})$ to be irreducible the separating line PQ must intersect the superior path but once. □

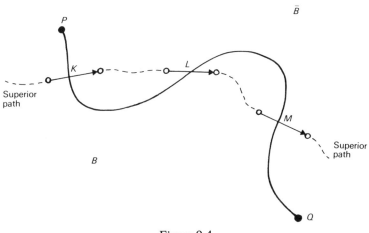

Figure 9.4

Theorem 9.4
The complete flow in standard network N, which results from applying the intuitive method to a geometrical representation without crossings, is a maximal flow.

Proof Let $(A, \bar{A})$ be any irreducible cut and PQ a separating line constructed as in the lemma.

At the first stage the largest possible flow is assigned to the superior path, the residual capacities updated and any saturated arcs dropped. $(A, \bar{A})$ remains irreducible and, by the lemma, has had its (residual) capacity reduced by the value of the flow assigned. The intuitive method repeats this process for as long as possible ending with a complete flow and with the (residual) capacity of $(A, \bar{A})$ having been reduced by the total value of the path flows assigned, that is by the value ϑ of the network flow.

Suppose now that B is the set of vertices which can be reached from S by a path consisting of unsaturated arcs only, and $\bar{B} = X - B$. $S \in B$ and $F \in \bar{B}$ and hence $(B, \bar{B})$ is a cut. Let $(D, \bar{D})$ be an irreducible reduction of $(B, \bar{B})$ then the residual capacity of $(D, \bar{D})$ is zero. Since $c(D, \bar{D})$ has also been reduced by ϑ it follows that $\vartheta = c(D, \bar{D})$ and hence the flow is maximal as required. □

The converse of this result is not true in that the intuitive method applied to geometrical representations with crossings still produces a maximal flow in some cases, (exercise 9.1).

9.2 ALGORITHMS FOR FINDING MAXIMAL FLOWS

We have seen that calculation of the partition $(A_f, \bar{A}_f)$ establishes the optimality of a flow f or provides a means of augmenting the flow. This approach forms the basis of the following algorithm which is essentially that in Ford and Fulkerson (1962).

Algorithm (FF)

Step 1 (Setup)
    Start with some initial feasible flow (which can be the zero flow with $f_{ij} = 0$ in all arcs ij).

Step 2 (Labelling)
    Assign the label $(+F)$ to S.
    Set $\text{scan}(i) = \textit{false}$ for all vertices i.
    Repeat step 2a until F is labelled or until $\text{scan}(i) = \textit{true}$ for all labelled i whichever is the sooner.

    Step 2a (Scan j)
        Select a labelled vertex j with $\text{scan}(j) = \textit{false}$.
        To each unlabelled successor k of j assign label $(+j)$ if jk is unsaturated.
        To each unlabelled predecessor i of j assign label $(-j)$ if ij is not flowless.
        Set $\text{scan}(j) = \textit{true}$.

Step 3 (Augmentation)
    If F is unlabelled terminate — a maximal flow has been found.
    Otherwise a flow augmenting chain is found by tracing back via the labels

to S and the flow augmented by as much as possible along this chain. Remove all labels and return to step 2.

Example 9.6 Use algorithm (FF) to find a maximal flow from source to sink starting with the flow of figure 9.5a.

Solution Label S $(+F)$ and set scan(i) = *false* all i. Scanning begins with vertex S the only labelled vertex. Assign label $(+S)$ to b and c as Sb and Sc are unsaturated.
Set scan (S) = *true*.
Scan c — this leads to no new labels.
Set scan(c) = *true*.
Scan b.
Assign label $(-b)$ to a as the flow in ab is non-zero.
Set scan(b) = *true*.
Scan a.
Assign label $(+a)$ to F.
Since F is labelled we have a flow augmenting chain $SbaF$ [the label of F is $(+a)$, the label of a is $(-b)$, the label of b is $(+S)$]. The flow is then augmented by 4 in the usual way (figure 9.5b).

The labelling phase starts again and it is found that this time only S, b and c can be labelled and so the flow is maximal.

That algorithm (FF) will produce an optimal solution if it terminates is clear from the results of the previous section. However Ford and Fulkerson (1962) have shown, by means of an example, that it is possible for it not to terminate if the capacities have irrational values. However, in practice the capacities will be integers or at worst rational numbers.

Theorem 9.5
If the capacities c_{ij} of a standard network are all rational numbers then there is a rational valued maximal flow with rational valued arc flows f_{ij}. Moreover this flow pattern can be obtained in a finite number of steps using algorithm (FF).

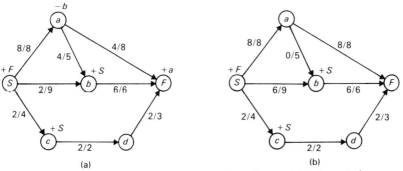

Figure 9.5 The numbers alongside each arc ij are of the form f_{ij}/c_{ij}

Proof Since the problem is linear we first convert all rational numbers to integers by multiplying all capacities by a suitable integer. The result then follows by applying algorithm (FF) starting with the zero flow as initial feasible flow. □

Although the conditions of the above theorem ensure termination the number of steps required can be arbitrarily large even for a small network (see exercise 9.2). The trouble lies in the way in which labelled vertex j is selected for scanning in step 2a of algorithm (FF). As it stands any choice is permissible. Two scanning strategies which could be used are breadth-first and depth-first (section 3.1). While DF might be useful for finding a complete flow, it could, at least in principle, cause trouble. On the other hand BF search does lead to flow augmenting chains containing a monotonically increasing number of arcs (Dinic, 1970, and Edmonds and Karp, 1972) and leads to theoretical bounds on the computational effort required in terms of the size (number of vertices) of the network.

If at some stage a feasible flow f has been constructed then the BF search can be performed using an algorithm of the following form which essentially corresponds to step 2 of algorithm (FF). Here X is the set of vertices of the network under consideration, and $d(i)$ is a label which gives the *depth* of vertex i from S and replaces the label scan(i).

Algorithm (BF)

Step 1 (Setup)
  Set
  $$d(i) = \begin{cases} 0 & \text{if } i = S \\ -1 & \text{otherwise} \end{cases}$$
  Set *current depth* = 1.

  Repeat step 2 while $d(F) = -1$.

Step 2 (Iteration)
  For all vertices $u \in X$ with $d(u) =$ *current depth* $- 1$:
  if $f_{uw} < c_{uw}$ or $f_{wu} > 0$ and $d(w) = -1$ then set $d(w) =$ *current depth* and set labels for flow calculations as appropriate.
  If no new vertices are labelled then the flow is maximal and the algorithm terminates.
  Current depth ← *current depth* + 1.

  Suppose $d(F) = k > 0$ and that $V = \bigcup_{\alpha=0}^{k} V^\alpha$ where

  $$V^\alpha = \{w \mid d(w) = \alpha\}$$

is the αth *layer*. Then the *DK augmentation network* (after Dinic and Karzanov) is the network DK(f) on vertex set V and with arcs ij such that

(1) $d(j) = d(i) + 1$ and $f_{ij} < c_{ij}$, or
(2) $d(i) = d(j) + 1$ and $f_{ij} > 0$.

Example 9.7 Find the DK augmentation network for the network of figure 9.1 with zero initial flow.

Solution It is easily checked that

$$V^0 = \{S\}, \quad V^1 = \{a, c, e\}, \quad V^2 = \{b, d, F\},$$

and that the DK augmentation network is as shown in figure 9.6a. Arcs, such as the ones shown by dashed lines in figure 9.6a from which F cannot be reached via vertices i with successively higher values of $d(i)$ are called *dead end*.

Theorem 9.6 (Monotonicity)
If a feasible flow $f = \{f_{ij}\}$ is augmented as much as possible along a shortest

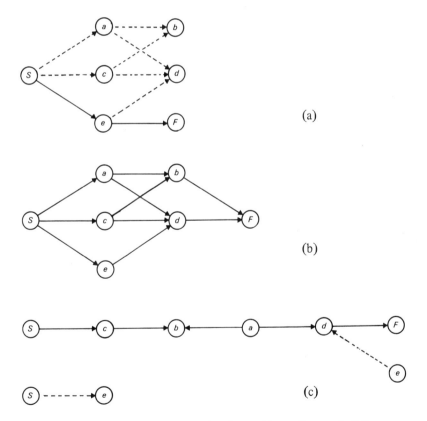

Figure 9.6 The DK network for the problem of example 9.7

(fewest arcs) flow augmenting chain to obtain a new feasible flow $f' = \{f'_{ij}\}$, then

$$d_{f'}(F) \geq d_f(F)$$

where the subscript denotes the flow with respect to which the labelling is performed.

Proof Let

$$\pi \equiv u_0 u_1 u_2 \ldots u_m$$

[where $u_0 = S$, $u_m = F$, $m = d_{f'}(F)$] be a shortest flow augmenting chain with respect to f'. Arcs of π which were arcs of DK(f) will be called *old* and the remainder will be called *new*. Let there be $N_{\text{old}}(\pi)$ and $N_{\text{new}}(\pi)$ of them respectively, then

$$d_{f'}(F) = N_{\text{old}}(\pi) + N_{\text{new}}(\pi)$$

Now for each forward (backward) old arc $u_{\alpha-1}u_\alpha$ ($u_\alpha u_{\alpha-1}$), $d_f(u_\alpha) = d_f(u_{\alpha-1}) + 1$. A new arc must have arisen by a reduction of flow in $u_{\alpha-1}u_\alpha$ (not $u_\alpha u_{\alpha-1}$ as this would imply we were dealing with an old arc) and so $d_f(u_\alpha) = d_f(u_{\alpha-1}) - 1$. As π only goes via vertices labelled with respect to f

$$d_f(F) \leq N_{\text{old}}(\pi) + N_{\text{new}}(\pi) = d_{f'}(F) \quad \square$$

Corollary 1 $d_{f'}(i) \geq d_f(i)$ for any $i \in \text{DK}(f) \cap \text{DK}(f')$.

Theorem 9.7
Algorithm (FF) with step 2 replaced by algorithm (BF) requires no more than $O(n^3)$ flow augmentations, where n is the number of vertices in the network.

Proof
(1)   There can be no more than p flow augmenting chains of a particular length d, where p is the number of arcs of the network.
   This is true since at each augmentation (at least) one arc becomes saturated or flowless. If this arc again lies on a flow augmenting chain it must be used in the opposite direction and as in theorem 9.6, the new chain must be longer than d. Thus the number of flow augmenting chains of length d clearly cannot exceed p.
(2) d can only take the values $1, 2, \ldots, n-1$ since a flow augmenting chain can pass through any vertex at most once.

Combining statements 1 and 2 shows that there cannot be more than $np < n^3$ augmentations. Moreover, Zadeh (1972) has identified a class of networks for which $O(n^3)$ augmentations is required. $\square$

Corollary   Algorithm (FF) with breadth first labelling requires no more than $O(n^5)$ computations.

Proof This is true since for each augmentation the most costly calculation is that of DK(f) but this cannot be worse than $O(n^2)$ since each arc can be examined at most twice. The result follows by using theorem 9.7. □

The above procedure is clearly wasteful as much information can be carried forward from one labelling to the next. It is modified by calculating just one DK augmentation network DK(f) for each length $d = d_f(F)$ of a flow augmenting chain, and removing dead ends.

At each augmentation along a chain of length d, $O(d)$ operations are required, and not more than $O(d)$ calculations to delete from DK(f) arcs which have become saturated or flowless. Since d is not greater than $n-1$, only $O(n)$ calculations at most are required per augmentation.

Theorem 9.8 (Dinic)
A maximal flow through a network can be found in not more than $O(n^4)$ operations.

Proof Since DK(f) is calculated only once for each depth $d = d_f(F)$ this leads to a total contribution (including the removal of dead ends) of not more than $O(n^3)$ operations.

On the other hand not more than np augmentations are required and so a total of not more than $O(n^2 p) \leqslant O(n^4)$ operations are required for the actual augmentations. The result follows. □

Example 9.8 Continue the solution started in example 9.7.

Solution The first DK augmentation network is given in figure 9.6a. Dead ends are removed leaving just the single path *Sef*. 3 units of flow are sent along *SeF*, *ef* becomes saturated and is removed, and there are no other flow augmenting chains of length 3.

DK(f) for the flow assigned is shown in figure 9.6b. There are no dead ends to be removed. 7 units of flow are sent along *SabF*. Arc *ab* becomes saturated and is removed. 1 unit of flow is sent along *SadF*, *Sa* becomes saturated and is removed. 1 unit of flow is sent along *ScbF*, *bF* becomes saturated and is removed, and further *cb* is now a dead end arc and is also removed. Next 3 units are sent via *ScdF*, *cd* becoming saturated and *Sc* dead end – both are removed. 5 units are sent via *SedF* and no flow augmenting chains of length 4 remain.

DK(f) for the flow so far assigned is shown in figure 9.6c. Notice that *ab* and *ed*, both of which are in figure 9.6b now appear but are used in opposite senses (see proof of theorem 9.7). 3 units of flow are sent along *ScbadF* and no flow augmenting chains of length 5 remain.

DK(f) for the flow thus far assigned does not contain F and hence the flow is maximal with value $(3+7+1+1+3+5+3) = 23$ units.

When finding maximal flows using algorithm (BF) the dominating term arises from the augmenting operations. Karzanov (1974) showed how this can be reduced by essentially augmenting all the flow augmenting chains of DK(f) at once in $O(n^2)$ operations, using the concept of pre-flows (see Malhotra et al., 1978). This leads to the following result.

Theorem 9.9 (Karzanov)
A maximal flow through a network can be found in $O(n^3)$ operations.

One further refinement may be mentioned. Pohl (1977) employed a bi-directional labelling scheme (compare section 4.1) with $d(i)$ as before and $db(i)$ the distance from F. Retaining those vertices i with minimal sum $d(i) + db(i)$ yields DK(f) with dead ends removed.

Cheung (1980) gives computational comparison of eight different methods for finding maximal flows. He found that for other than small networks Dinic's algorithm was the fastest [though only slightly faster than using the BF version of algorithm (FF) or using pre-flows]. Specimen average times using Dinic's algorithm implemented in Algol 60 on an IBM 360/65 computer are

(1) 0.45 seconds for 50 vertex 246 arc (density = 0.1) networks;
(2) 10.38 seconds for 200 vertex 1900 arc (density = 0.05) networks;
(3) 203.63 seconds for 1500 vertex 5622 arc (density = 0.0025) networks.

EXERCISES

9.1 Let N be the network obtained from that of example 9.1, with the capacity of arc cb being reduced to 1 (all other capacities remaining the same). Prove that although this is a non-planar network, the method of assigning as much flow as possible to paths in order of decreasing superiority leads to a complete flow which is also maximal.

9.2 Apply algorithm (FF) to the network in figure 9.7 with (a) lowest numbered vertices, and (b) highest numbered vertices selected at alternate applications of step 2. Verify that 2000 flow augmentations are required.

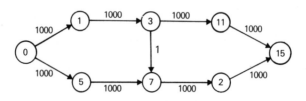

Figure 9.7 The number alongside arc ij is the capacity c_{ij}

9.3 Use the Ford–Fulkerson labelling algorithm [algorithm (FF)] to find a maximal flow from S to F through the network in figure 9.8, starting from the initial flow given. The numbers alongside arc xy are of the form c_{xy}/f_{xy} where c_{xy} is the capacity of xy and f_{xy} is the flow along xy.

(LU 1978, abridged.)

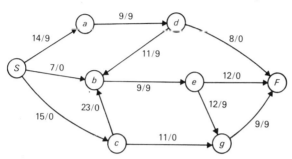

Figure 9.8

9.4 A Housing Authority classifies tenants of dwellings under its control into $n+1$ categories. Category 0 comprises tenants who are urgently in need of being rehoused. Categories 1 to n relate to the type of dwelling presently occupied.

Let c_{ij} be the number of tenants wishing to move from a type i house to a type j house ($i \neq j \in \{1, \ldots, n\}$) and let $c_{i,n+1}$ be the number of vacancies in type i houses. The policy of the Authority is to authorise such moves as will result in as many as possible of the category 0 tenants being rehoused. Model this as a maximal flow problem.

Find how many category 0 tenants can be rehoused if $n = 4$ and the matrix (c_{ij}) is

	0	1	2	3	4	5
0	0	20	15	17	12	0
1	∞	0	9	1	3	6
2	∞	11	0	2	12	18
3	∞	4	2	0	2	15
4	∞	0	10	3	0	22

10 Network Flow: Extensions

The basic problem studied in chapter 9 was indeed very specialised and many features that might be expected in a real life situation were not accounted for. This chapter extends the basic theory in several directions. Section 10.1 deals with dynamic (or time varying) flows, non-conserved flows and multi-commodity flows as well as some simple modifications to the basic theory. Section 10.2 centres around the concept of least cost flow of a particular value (which may be maximal); for the most part the models allow only cost functions linear in arc flow values, but non-linear cost functions are treated briefly. The third section deals with the efficient application of the simplex method to network problems.

10.1 VARIOUS EXTENSIONS

(a) Undirected and Mixed Networks

So far only flows in directed networks have been considered. However, flows in undirected and mixed networks can be brought within the same framework merely by replacing each edge ij by a pair of arcs ij and ji, and replacing the edge flow constraint $-k \leqslant f_{ij} \leqslant k$ by the two arc flow constraints $0 \leqslant f_{ij} \leqslant k$, $0 \leqslant f_{ji} \leqslant k$.

(b) Multiple Sources and/or Sinks

A network N with several sources $S_1, S_2, \ldots, S_p$, from which unlimited flow can originate, can be converted to a network with a single source S by introducing vertex S (which will be called a *super source*) together with arcs $SS_1, SS_2, \ldots, SS_p$ each with infinite capacity. On the other hand if the flow in N originating at S_i is limited, to not more than a_i say, then this can be accounted for by setting the capacity of SS_i to be a_i. Similar remarks apply if N has more than one sink.

Example 10.1 (Transportation problem)   A company has three factories P, Q and R with weekly requirements of 5 tonnes, 9 tonnes and 6 tonnes respectively, of a particular raw material which may be obtained from three suppliers A, B and C. Suppliers A and B can each supply up to 8 tonnes at costs of £100 per tonne and £105 per tonne respectively. Supplier C is prepared to meet any demand that may be made but at a cost of £120 per tonne. In view of this higher price the company decides to acquire as little as possible of the raw material from C subject only to the condition that all demands must be met. Transport costs are assumed to be linear and fixed at £1.5 per tonne per km carried. The distances (in km) between supply and demand points are given in the table below. In view of the relatively high distances involved it is decided to avoid routes A to R and C to P. Express the problem of meeting all demands at minimal cost as a network flow problem.

	P	Q	R
A	12	14	32
B	18	22	14
C	30	14	10

Solution First note that the company's decision concerning supply from C is equivalent to C imposing a limit of $(5+9+6-8-8)=4$ tonnes per week on the amount of raw material C will supply. The problem is now a flow problem with three 'capacitated' sources A, B and C and three 'capacitated' sinks P, Q and R. Adding a super source S and super sink F and inserting appropriate arc capacities and transport rates, the problem is that of finding a least cost maximal flow from S to F through the network of figure 10.1. Using the classical transportation algorithm it is easily verified that an optimal solution to the problem is to supply each week

3 tonnes to P from A
5 tonnes to Q from A
2 tonnes to P from B
6 tonnes to R from B
4 tonnes to Q from C

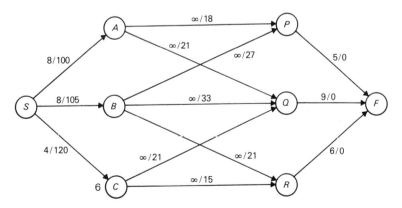

Figure 10.1 Numbers alongside arcs are of the form capacity/cost per unit flow

In section 10.2 it is shown how this problem could be solved as a minimal cost network flow problem.

(c) Absence of Source and/or Sink

Suppose that all flow originates at a vertex S, which has an incoming arc bS, and terminates at a vertex F.

Theorem 10.1
There exists a maximal flow f from S to F with $f_{bS} = 0$ for any arc bS incoming at S.

Proof Suppose $f^0_{bS} > 0$ for some maximal flow f^0. Then, since all flow must originate at S, bS must be part of a circuit $\sigma \equiv Su_1u_2\ldots u_pbS$ with $f^0_{xy} > 0$ for each arc xy in σ. Reducing the flow in each arc $xy \in \sigma$ by $k = \min_{xy \in \sigma} f_{xy}$ leads to a new flow $f' = \{f'_{xy}\}$ with $\vartheta_{f'} = \vartheta_f$.

If $f'_{bS} = 0$ the proof is complete. Otherwise there is some other circuit σ' containing bS for which $f'_{xy} > 0$ all $xy \in \sigma'$. The flow around σ' can now be reduced by $\min_{xy \in \sigma'} f'_{xy}$ to obtain a new flow f'' with $\vartheta_{f''} = \vartheta_{f'} = \vartheta_f$. This procedure of reducing flows round circuits containing bS continues until the flow in bS has been reduced to zero. This must eventually happen since we deal only with finite networks and at least one new arc becomes flowless at each flow reduction. □

Corollary 1 There exists a maximal flow f for which every arc incoming to S is flowless.

Corollary 2 The maximum flow value is unaltered if all arcs incoming to S and outgoing from F are deleted.

The above result, together with the modifications discussed in (b), shows that nothing essential is lost by assuming a unique source at which flow originates and a unique sink at which flow terminates.

(d) Vertex Constraints

It is probably true that most motor traffic hold-ups occur because a junction (roundabout, intersection with or without traffic lights) is unable to cope with all the vehicles that require to use it. Thus it is clear that, for some problems at least, it is important to take account of constraints on flow through vertices.

Suppose that the total flow f_b through vertex b is limited by $0 \leqslant f_b \leqslant c_b$ where c_b is the vertex capacity of b. These restrictions can be expressed in terms of arc flows f_{ij} by

$$\sum_{i \in \Gamma^{-1}b} f_{ib} \leqslant c_b \qquad \sum_{j \in \Gamma b} f_{bj} \leqslant c_b$$

Network Flow: Extensions   217

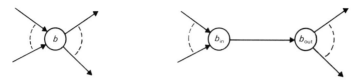

Figure 10.2

But these are equivalent to the constraints that would be obtained if b is replaced by two uncapacitated vertices b_{in} and b_{out} connected by a single arc from b_{in} to b_{out} of capacity c_b, and such that all arcs incoming to b become arcs incoming at b_{in} and all arcs outgoing from b become arcs outgoing from b_{out} (see figure 10.2). Thus from the point of view of finding maximal network flows it is sufficient to replace each capacitated vertex by two uncapacitated vertices connected by a capacitated arc.

Example 10.2 Part of a one-way traffic system is shown in figure 10.3a with the capacities and permitted directions of flow for each road section. Also junctions A and B have capacities of 17 and 12 respectively while junction C may be assumed to impose no constraints on any possible flows. What is the maximal steady flow through the system and what modification to the permitted directions of flow is advisable?

(LU 1971, abridged)

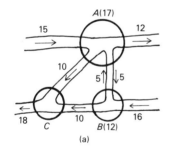
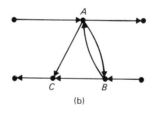
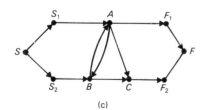
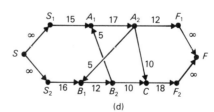

Figure 10.3

Solution The network corresponding to the traffic problem is shown in figure 10.3b and, with a super source S and super sink F added, in figure 10.3c. Vertices A and B are split into vertices A_1, A_2 and B_1, B_2 respectively with the capacities of $A_1 A_2$ and $B_1 B_2$ being set at 17 and 12 respectively. The problem has now been converted to that of finding a maximal flow through the standard network shown in figure 10.3d.

It is easily verified that the assignments

12 units to $SS_1 A_1 A_2 F_1 F$
3 units to $SS_1 A_1 A_2 CF_2 F$
10 units to $SS_2 B_1 B_2 CF_2 F$
2 units to $SS_2 B_1 B_2 A_1 A_2 CF_2 F$

lead to a (maximal) flow of value 27. Thus even if the above assignments reflect driver intentions the flow cannot exceed 27 and may well be less in practice.

There are just two minimal cuts (or capacity 27) namely, ($\{S, S_1, S_2, B_1\}$, $\{A_1, A_2, B_2, C, F_1, F_2, F\}$) and ($\{S, S_1, S_2, A_1, B_1, B_2\}, \{A_2, C, F_1, F_2, F\}$). Since arc $A_2 B_1$ is a *reverse* arc, (that is, 'passes from the sink side of the cut to the source side') it will be flowless for any maximal flow. Consequently it would seem preferable, as maximum throughput is being aimed for, to restrict flow passing between A and B to the direction BA. That is, the corresponding road section should be made one-way in the direction BA.

(e) Lower Bound Constraints

It sometimes happens that the rate of flow through a particular component of a network must exceed a certain critical value. For example, the flow rate of soil sewage through pipes must exceed a minimal cleansing velocity to avoid silting up; too low a flow rate of a molten chemical through pipes might be intolerable due to the possibility of solidification.

To incorporate such lower bounds the constraints 9.2b and c of the basic problem are modified to

$$0 \leq l_{ij} \leq f_{ij} \leq c_{ij} \qquad (10.1)$$

where l_{ij} and c_{ij} are constants. This leads to the following variant of the max-flow min-cut theorem (theorem 9.3) in which, given any set $\{\psi_{ij}\}$ of numbers defined on the arc set, $\psi(X, Y)$ will denote

$$\psi(X, Y) = \sum_{\substack{x \in X \\ y \in Y}} \psi_{xy}$$

for any proper partition (X, Y) of the set of vertices.

Theorem 10.2
If any feasible flow exists through a network N then

$$\max_f \vartheta_f = \min_{(A,\bar{A})} [c(A,\bar{A}) - l(\bar{A},A)]$$

where $(A,\bar{A})$ is a cut.

Proof Clearly (see theorem 9.1)

$$\max_f \vartheta_f \leq \min_{(A,\bar{A})} \left\{ \sum_{\substack{i \in A \\ j \in \bar{A}}} c_{ij} - \sum_{\substack{u \in \bar{A} \\ w \in A}} l_{uw} \right\}$$

That is

$$\max_f \vartheta_f \leq \min_{(A,\bar{A})} [c(A,\bar{A}) - l(\bar{A},A)] \tag{10.2}$$

It remains to be shown that the strict inequality cannot hold. To do this a partition $(L_f, \bar{L}_f)$, corresponding to a flow f, is formed as follows

(1) $S \in L_f$
(2) If $j \in L_f$ and $f_{jk} < c_{jk}$ add k to L_f
(3) If $j \in L_f$ and $f_{ij} < l_{ij}$ add i to L_f
(4) $\bar{L}_f = X - L_f$

Suppose $F \in L_f$, then, just as in the proof of theorem 9.2, a flow augmenting chain from S to F exists and the value of the flow may be increased by a (strictly) positive amount. Consequently for a maximal flow f^*, F must belong to $\bar{L}_{f^*}$ and $(L_{f^*}, \bar{L}_{f^*})$ is a cut. The value of the flow is then the net value of the flow across this cut

$$\vartheta_{f^*} = \sum_{\substack{i \in L_{f^*} \\ j \in \bar{L}_{f^*}}} c_{ij} - \sum_{\substack{u \in \bar{L}_{f^*} \\ w \in L_{f^*}}} l_{uw}$$

$$= c(L_{f^*}, \bar{L}_{f^*}) - l(\bar{L}_{f^*}, L_{f^*})$$

$$\geq \min_{(A,\bar{A})} [c(A,\bar{A}) - l(\bar{A},A)]$$

The result follows immediately by comparing this inequality with that of inequality 10.2. □

Since lower limits, $l_{ij} > 0$, to the values of arc flows f_{ij} are permitted it is likely that the minimum value of the network flow will be non-zero.

Theorem 10.3
If any feasible flow exists through a network N then

$$\min_f \vartheta_f = \max_{(A,\bar{A})} \left\{ l(A,\bar{A}) - c(\bar{A},A) \right\}$$

Proof (See exercise 10.2.)

Notice that $\min_f \vartheta_f$ must be non-negative since flow is only allowed to originate at S. Further the proof of theorem 10.2 relies on the definition of the partition $(L_f, \bar{L}_f)$ which in turn relies on the existence of a feasible flow f. (Similar remarks may be made about the proof of theorem 10.3.) How can feasible flows be found, if any exist, and under what circumstances will one fail to exist?

In order to answer this question let f be any flow that would be feasible in the *absence* of lower bound constraints (that is, $0 \leqslant f_{ij} \leqslant c_{ij}$ all ij). Then the *kilter number*

$$K_{ij} = \max(0, l_{ij} - f_{ij})$$

is the amount by which f_{ij} fails to be a feasible value for a flow in arc ij. If $K_{ij} > 0$ the arc ij is *out of kilter*; otherwise it is *in kilter*.

Let pq be any arc with $K_{pq} > 0$ and define a network $\tilde{N}(pq)$ which differs from N only in that arc pq is removed and an arc FS added with $l_{FS} = 0$, $c_{FS} = \infty$. A flow $f = \{f_{ij}\}$ in N corresponds to a flow $\tilde{f} = \{\tilde{f}_{ij}\}$ in $\tilde{N}(pq)$ from q to p, where

$$\tilde{f}_{ij} = f_{ij} \qquad ij \neq pq, FS \qquad (10.3)$$

$$\tilde{f}_{FS} = \vartheta_f$$

$\tilde{f}$ is clearly a conserved flow in $\tilde{N}(pq)$. Flow through $\tilde{N}(pq)$ is now augmented using a labelling algorithm which differs from that of algorithm (FF) (see section 9.2) only in that

'label $(-j)$ if ij is not flowless'

is replaced in step 2a by

'label $(-j)$ if $\tilde{f}_{ij} > l_{ij}$'

and in step 3 the augmentation ϵ must not exceed $f_{ij} - l_{ij}$ on backward arcs.

Relating this to network N we have, via equation 10.3, a new flow f' for which K_{pq} is decreased and no kilter number K_{ij} is increased. The augmentations are continued until $K_{pq} = 0$ and thus there is (at least) one more arc in kilter. The procedure is now repeated for another arc still out of kilter, and so on until a feasible flow is attained. This procedure only breaks down if, after a labelling step, the set of labelled vertices Q_f does not include p. Then in N

$$f_{ij} = c_{ij} \qquad i \in Q_f, \; j \in \bar{Q}_f$$

$$f_{uw} \leqslant l_{uw} \qquad u \in \bar{Q}_f, \; w \in Q_f$$

with the strict inequality $f_{pq} < l_{pq}$ holding. Summing gives

$$c(Q_f, \bar{Q}_f) = \sum_{\substack{i \in Q_f \\ j \in \bar{Q}_f}} c_{ij} = \sum_{\substack{i \in Q_f \\ j \in \bar{Q}_f}} f_{ij} = \sum_{\substack{u \in \bar{Q}_f \\ w \in Q_f}} f_{uw} < \sum_{\substack{u \in \bar{Q}_f \\ w \in Q_f}} l_{uw} = l(\bar{Q}_f, Q_f)$$

Theorem 10.4
A feasible flow satisfying lower bound constraints exists in a network N if and only if, for all cuts $(A, \bar{A})$

$$c(A, \bar{A}) - l(\bar{A}, A) \geq 0 \qquad (10.4)$$

Proof The above construction shows how a feasible flow is achieved if equation 10.4 holds. The construction is certainly finite if all l_{ij} and c_{ij} and the initial f_{ij} are rational numbers, as is assumed to be the case.
  Conversely if equation 10.4 does not hold there is clearly no feasible flow. □

The related networks were introduced in order to simplify the proof. However, as shown by the following example, they can be dispensed with in practice if so desired.

Example 10.3 Let f be the flow through the network of figure 10.4a obtained by assigning flows of 3, 4, 4 and 2 to paths $SbcF$, $SbcdF$, $SadF$ and $SaeF$ respectively. Is f feasible? If not, then use it to find a flow that is feasible.

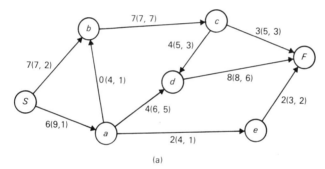

(a)

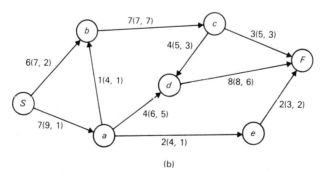

(b)

Figure 10.4 Numbers alongside arcs are of the form flow (capacity, lower bound)

Solution Clearly flow is conserved. Also the capacity constraints $f_{ij} \leq c_{ij}$ are satisfied for all arcs ij. However, the lower bound constraints are violated for arcs ab, ad; the associated kilter numbers are

$$K_{ab} = l_{ab} - f_{ab} = 1 - 0 = 1$$
$$K_{ad} = l_{ad} - f_{ad} = 5 - 4 = 1$$

Consider arc ab first. Assign label $(+a)$ to b. Then S is labelled $(-b)$. Vertex a can now be labelled $(+S)$ and breakthrough has been achieved. A flow of

$$\min(c_{ab} - f_{ab}, f_{Sb} - l_{Sb}, c_{Sa} - f_{Sa}) = 1$$

may be assigned to the cycle $abSa$. This is sufficient to bring arc ab into kilter (figure 10.4b).

Only arc ad remains out of kilter. Assign label $(+a)$ to d. Then c is labelled $(-d)$, then F can be labelled $(+c)$. [Note that b cannot be labelled $(-c)$ even though it is not flowless.] At this point we have broken through to F and $\min(f_{cd} - l_{cd}, c_{cF} - f_{cF}) = 1$ unit of flow may be assigned to chain dcF. Now labelling continues from S. Assign label $(+F)$ to S. Both a and b may be labelled $(+S)$. We have broken through to a and $\min(c_{Sa} - f_{Sa}) = 2$ units of flow may be assigned to chain Sa. Combining the three chains Sa, ad and dcF it is seen that

$$\min(2, c_{ad} - f_{ad}, 1) = 1$$

unit of flow may be assigned to chain $SadcF$ without making any in kilter arc out-of-kilter. Also this assignment is sufficient to bring arc ad into kilter and a feasible flow has been achieved. (Is it minimal, maximal or neither?)

(f) Dynamic Flows

Static (or non-varying) flows are of relevance in situations in which the flow is maintained in an essentially steady state over a long period of time. What happens, however, if the rate of flow is variable or only a short time period is being considered? Clearly the maximum value of a dynamic flow cannot be less than the value of a maximal static flow, but can it be greater? These questions will be answered by an example which also shows how the theory of dynamic flow through a network can be represented as that of a static flow through a modified network.

Example 10.4 Traffic enters the network in figure 10.5 at A and leaves at D. The numbers alongside each arc ij are of the form c_{ij}/t_{ij} where c_{ij} is the capacity of arc ij and t_{ij} is the time required for flow to pass from i to j via ij. What is the maximum amount of flow that can arrive at D in a unit time interval? What is the maximum amount that can arrive in three time intervals?

Network Flow: Extensions 223

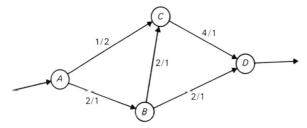

Figure 10.5

Solution The maximum steady flow that can be maintained indefinitely, that is the maximal static flow, is seen to have value 3, ($\{A\}$, $\{B, C, D\}$) being a minimal cut.

Suppose now that

at time 0	2 units of flow leave A along AB
	1 unit of flow leaves A along AC
at time 1	2 units of flow leave A along AB
	2 units of flow will arrive at B from A and are directed along BC
at time 2	2 units of flow arrive at B from A and are directed along BD
	3 units of flow arrive at C (2 from B and 1 from A) and are directed along CD
at time 3	5 units of flow arrive at D

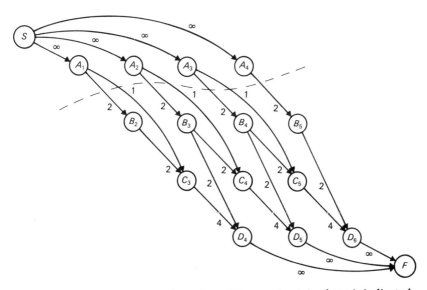

Figure 10.6 Capacities are given alongside arcs. A minimal cut is indicated

Thus, although the value of a maximal static flow is 3, as many as 5 units of flow may arrive in a single time period if the flow varies. In order to study this problem further a new network is formed whose vertices are A_t, B_t, C_t, and D_t where X_t denotes vertex X at time t. There is an arc $X_t Y_{t+\tau}$ of capacity c_{XY} only if $\tau = t_{XY}$. Adding a super source and a super sink the network of figure 10.6 is obtained. Notice that vertices B_1, C_1, C_2, A_5, C_6, etc., have not been added as flow through these vertices could not reach D in the required period of three intervals (that is, could not reach D_4, D_5 or D_6). Clearly a maximal static flow through this modified network is equivalent to a dynamic flow for which a maximal flow, over three time periods, is registered at D.

It may be shown that 11 units is the maximal amount of flow that can arrive at D in three consecutive time periods.

(g) Flows with Gains

Until now it has been assumed that any flow entering an arc ij at i will leave at j. However flow may be lost, for example, by leakage from electric power lines, by wastage of material in manufacturing processes, by loss of information in noisy communication channels. On the other hand flow may be gained; for example, value is added as items go through a manufacturing process. To take account of this arc flows f_{ij} are replaced by a pair of flows: the *entering* flow f^E_{ij} and the *leaving* flow f^L_{ij}. The two flows are related by $f^L_{ij} = g_{ij} f^E_{ij}$ where $g_{ij} > 0$ is termed the *gain* (though an actual loss of flow is implied if $g_{ij} < 1$). It is assumed that flow is still conserved at vertices and that there is a limit on the amount of flow in any arc. Then the following maximal flow problem can be posed, where ϑ_S and ϑ_F denote respectively the total flow out of S and the total flow into F

$$\text{maximise} \quad \vartheta_F = \sum_{i \in \Gamma^{-1} F} f_{iF}$$

subject to

$$f^L_{ij} = g_{ij} f^E_{ij}$$

$$0 \leq f^E_{ij} \leq c_{ij}$$

$$\sum_{i \in \Gamma^{-1} j} f^L_{ij} - \sum_{k \in \Gamma j} f^E_{jk} = \begin{cases} 0, & j \neq S, F \\ -\vartheta_S, & j = S \\ \vartheta_F, & j = F \end{cases}$$

Notice that, unlike the case in section 9.1, the conservation equations form an independent set. A chain

$$\pi \equiv \underset{\substack{\| \\ S}}{u_0} u_1 u_2 \ldots \underset{\substack{\| \\ F}}{u_p}$$

is flow augmenting, if flow can be sent along it from S to F, that is, if

$$f^E_{u_\alpha u_{\alpha+1}} < c_{u_\alpha u_{\alpha+1}} \quad \text{if } u_\alpha u_{\alpha+1} \in FA(\pi)$$
$$f^E_{u_{\alpha+1} u_\alpha} > 0 \quad \text{if } u_{\alpha+1} u_\alpha \in BA(\pi)$$

where $FA(\pi)$ is the set of forward arcs of π (relative to the given flow) and $BA(\pi)$ is the set of backward arcs of π. (See p. 202 for definition.) By a straightforward modification of algorithm (FF) a maximal flow can be found in networks without circuits, and the following result obtained. (The proof is left as an exercise.)

Theorem 10.5
A flow in a network with gains but possessing no circuits is maximal if there is a saturated cut (that is, the flow across the cut in the forward direction is equal to the capacity of the cut).

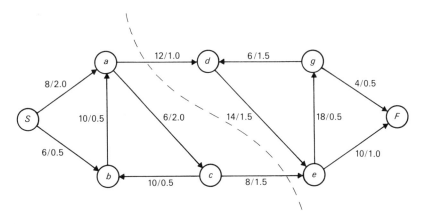

Figure 10.7 Numbers alongside each arc are of the form c_{ij}/g_{ij}

Readers may have wondered why the condition that the network has no circuits should be introduced. However, this was necessary since flow may be 'created' or 'absorbed', round circuits as the following example shows.

Example 10.5 Show that there are feasible flows in the network of figure 10.7 such that
(1) $\vartheta_S > 0, \vartheta_F = 0$;
(2) $\vartheta_S = 0, \vartheta_F > 0$.

Solution (1) Let flow in all arcs be zero except for the following

$f^E_{Sa} = 1$ $\quad\quad$ $f^L_{Sa} = 2$

$f^E_{ac} = 4$ $\quad\quad$ $f^L_{ac} = 8$

$f^E_{cb} = 8$ $\quad\quad$ $f^L_{cb} = 4$

$f^E_{ba} = 4$ $\quad\quad$ $f^L_{ba} = 2$

$\vartheta_S = 1, \vartheta_F = 0$.

(2) Let all flows be zero except for the following

$f^E_{gd} = 6$ $\quad\quad$ $f^L_{gd} = 9$

$f^E_{de} = 9$ $\quad\quad$ $f^L_{de} = 13.5$

$f^E_{eg} = 12$ $\quad\quad$ $f^L_{eg} = 6$

$f^E_{eF} = 1.5$ $\quad\quad$ $f^L_{eF} = 1.5$.

$\vartheta_S = 0, \vartheta_F = 1.5$.

It is interesting to note that the flows of (1) and (2) can be combined to give a network flow with $\vartheta_S = 1$, $\vartheta_F = 1.5$ such that no flow crosses the cut marked in figure 10.7.

The ideas of 'absorbing' and 'creating' can be extended to cycles. (Note that if the flow round a cycle is capable of being increased or decreased then the cycle is absorbing for one orientation and creating for the other.) An absorbing cycle is *active* if its flow can be augmented (for the direction in which the cycle is absorbing) and flow can be sent from S to some point on the cycle. The cycle *acba* is active for the flow in the solution of example 10.5 part (1). If an active absorbing cycle exists then ϑ_S can be increased without increasing ϑ_F.

Similarly a creating cycle is active if its flow can be augmented and flow can be sent from some point on the cycle to F. If an active creating cycle exists then ϑ_F can be increased without altering ϑ_S. The cycle *degd* for the flow in the solution of example 10.5 part (2) is inactive since arc *gd* is saturated. However, the cycle *abca* (that is, the circuit *acba* in the reverse direction) is creating and active. For example $f^E_{ba}, f^E_{ac}, f^E_{cb}$, can be reduced by 4, 2 and 8 respectively and f^E_{ce}, f^E_{eF} increased by 4 and 6 respectively. Thus ϑ_F has been increased by 6 without altering ϑ_S and since the flow in *cb* and *ba* cannot be further reduced *abca* is now inactive as a creating cycle. The following result is readily established and leads to an algorithm for finding maximal flows (exercise 10.2).

Theorem 10.6

A flow in a network with gains is maximal if there is a saturated cut and no active creating cycles.

(h) Multi-commodity Flows

In transport and communication systems it is likely that more than one commodity will be carried by the network. Different commodities may correspond to different types of goods or to one type of goods which must flow between different pairs of vertices. (For example, in the U.K. telephone system, calls from Bradford to Bristol must be regarded as being a different commodity from calls from Cardiff to Leeds, or even calls from Bristol to Bradford.) The essential feature of multi-commodity flows is that flows in opposite directions of two different commodities do *not* cancel.

Denote by f^α_{ij} the flow of the αth commodity along arc ij, and by S^α and F^α the source and sink for the αth commodity. The network flow $f = \{f^\alpha_{ij}\}$ is a feasible multi-commodity flow if, for all α

$$\sum_{i \in \Gamma^{-1}j} f^\alpha_{ij} - \sum_{k \in \Gamma j} f^\alpha_{jk} = \begin{cases} -\vartheta^\alpha & \text{if } j = S^\alpha \\ 0 & \text{if } j \neq S^\alpha, F^\alpha \\ \vartheta^\alpha & \text{if } j = F^\alpha \end{cases} \quad (10.5)$$

$$\sum_\alpha f^\alpha_{ij} \leq c_{ij} \quad \text{for all } ij \quad (10.6)$$

$$f^\alpha_{ij} \geq 0 \quad \text{for all } ij \quad (10.7)$$

Constraints 10.5 to 7 are clearly the generalisations of the conservation, capacity and non-negativity constraints 9.2 (p.197) for single commodity flow. Notice that if constraint 10.6 were to be replaced by '$f^\alpha_{ij} \leq c^\alpha_{ij}$ for all α' then the problem would simplify to a set of single commodity problems with one for each α.

In the maximal multi-commodity flow problem it is required that $\vartheta_f = \sum_\alpha \vartheta^\alpha$ be maximised. It is immediately clear that

$$\max_f \vartheta_f \leq \sum_\alpha \max \vartheta^\alpha$$

The strict inequality may hold. Also, if c_{ij} are all integers it is not necessarily true that there exists a maximal multi-commodity flow with f^α_{ij} integer or even ϑ^α integer (compare theorem 9.5). These points are illustrated by the following example.

Example 10.6 Three commodities can flow in a network N on vertex set $\{1,2,3\}$ and arc set $\{12, 23, 31\}$.

For commodity 1 the source is vertex 1 and the sink is vertex 3;
For commodity 2 the source is vertex 2 and the sink is vertex 1;
For commodity 3 the source is vertex 3 and the sink is vertex 2.

Determine a maximal multi-commodity flow if the (total) capacity of each arc of N is 1.

Solution max ϑ^1 = max ϑ^2 = max ϑ^3 = 1. However if ϑ^α = 1 then ϑ^β must be zero for $\beta \neq \alpha$, and two arcs are saturated and one flowless. A more promising flow in which all three arcs are used to capacity is

$$f_{12}^1 = f_{23}^1 = f_{23}^2 = f_{31}^2 = f_{31}^3 = f_{12}^3 = \tfrac{1}{2}, \qquad f_{31}^1 = f_{12}^2 = f_{23}^3 = 0$$

for which $\vartheta^1 = \vartheta^2 = \vartheta^3 = \tfrac{1}{2}$ and $\vartheta_f = \tfrac{3}{2}$. Is this flow maximal? The answer is 'yes' as can be seen from the fact that all flow uses exactly 2 arcs and all arcs are used to capacity.

The general problem of finding maximal multi-commodity flows is rather difficult. Although upper bounds may be obtained by generalising the ideas of a cut the bounds are not always attained. Also multi-commodity flow in undirected networks cannot be treated simply by replacing each edge by a pair of oppositely directed arcs. For a further discussion of multi-commodity flows the reader is referred to Hu (1969) and Frank and Frisch (1971).

10.2 MINIMAL COST FLOWS

The discussion so far has related to finding maximal flows or, when lower bound constraints are present, to finding feasible flows. However, in practice a minimal cost flow of a particular value (whether maximal or not) is often of major interest. This is the case for the transportation problem (example 10.1) and for the problem below.

Example 10.7 [Assignment of pupils to schools (see Belford and Ratliff, 1972)] Two ethnic groups, which we denote by G and H, are predominant in a certain city. It is the policy of the authorities that the proportion of G (or H) in each school should be approximately the same as that for the city as a whole. Belford and Ratliff discuss such a 'balance' requirement in relation to U.S. legislation. Let

c_j = total number of pupils assigned to school j
p_i^G (p_i^H) = number of pupils in area i of group $G(H)$
f_{ij}^G (f_{ij}^H) = number of pupils of group G (H) from area i attending school j
D_j^G (D_j^H) = desired values of the total number of pupils of group G (H) attending school j
e_j^G (e_j^H) = excess pupils of group G (H) attending school j
d_{ij} = distance between area i and school j

The balance requirement can reasonably be translated into upper bounds U_j^G (U_j^H) on the numbers of pupils of group G (H) who may attend school j. The desired number of pupils of group G (H) attending school j could be such that the balance of the groups in the school reflects the balance in the neighbourhood of the school. However, D^G_j (D^H_j) can be any numbers for which the upper bound constraints are not violated and for which $D^G_j + D^H_j = c_j$ for all j. Then

$$e^G_j = \max\left(\sum_i f^G_{ij} - D^G_j, 0\right),\ e^H_j = \max\left(\sum_i f^H_{ij} - D^H_j, 0\right)$$

The problem of minimising the total number of pupil-kilometres travelled per day becomes

$$\text{minimise} \quad \sum_{i,j} d_{ij}(f^G_{ij} + f^H_{ij}) \tag{10.8}$$

subject to

$$\left.\begin{array}{l} \sum_j f^G_{ij} = p^G_i \\[1ex] \sum_j f^H_{ij} = p^H_i \end{array}\right\} \tag{10.9}$$

$$\sum_i f^G_{ij} - e^G_j + e^H_j = D^G_j \tag{10.10a}$$

$$\sum_i f^H_{ij} - e^H_j + e^G_j = D^H_j \tag{10.10b}$$

$$\sum_i (f^G_{ij} + f^H_{ij}) = c_j \tag{10.10c}$$

$$e^G_j \leqslant U^G_j - D^G_j \tag{10.11a}$$

$$e^H_j \leqslant U^H_j - D^H_j \tag{10.11b}$$

$$e^G_j, e^H_j, f^G_{ij}, f^H_{ij} \geqslant 0 \quad \text{for all } i, j$$

This can be interpreted as a minimal cost maximal flow problem as follows. The vertices of the network are S (the source), F (the sink), a^G_i, a^H_i corresponding to area i and groups G, H, and s^G_j, s^H_j corresponding to school j and groups G, H (see figure 10.8). f^G_{ij} (f^H_{ij}) is the flow from a^G_i to s^G_j (a^H_i to s^H_j) and e^G_j (e^H_j) is the flow from s^G_j to s^H_j (s^H_j to s^G_j). The constraints of equations 10.10a to c are conservation constraints at vertices s^G_j and s^H_j, and the constraints of equations 10.9 are conservation constraints at vertices a^G_i, a^H_i. Finally the constraints 10.11 can be represented as capacity constraints on arcs Sa^G_i, Sa^H_i, $s^G_j F$, $s^H_j F$ together with insistence on a maximal flow [of value $\sum_i (p^G_i + p^H_i)$].

Since the objective in 10.8 is just the total number of pupil-kilometres travelled, or cost of the flow, we clearly have a minimal cost maximal flow problem. (Figure 10.8 illustrates the problem for the special, very simple, case of 2 schools and 3 pupil areas.)

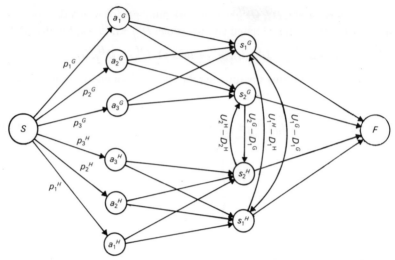

Figure 10.8 Numbers alongside arcs are capacities. Where not explicitly given capacities may be assumed to be ∞

Belford and Ratliff applied their model to the Gainesville, Florida, school system. They also discussed various refinements to the model. In particular the desirability of balance could also be represented by costs on the arcs $s_j^G s_j^H$ and $s_j^H s_j^G$; indeed if these costs are sufficiently large then the constraints 10.11 a and b could be dispensed with, but it is not clear that this is a desirable approach to adopt.

In the above example no lower bound constraints were present and for problems of this type there are two simple algorithms, for finding minimal cost flows with a particular value, which depend on the theorems below.

A *length* $l(\pi)$ is defined for each chain $\pi \equiv u_0 u_1 \ldots u_m$

$$l(\pi) = \sum_{FA(\pi)} p_{u_\alpha u_{\alpha+1}} - \sum_{BA(\pi)} p_{u_{\alpha+1} u_\alpha}$$

where p_{ij} is the length of arc ij. Also the term 'flow augmenting' is applied to cycles ρ which permit the addition of an extra positive flow in the direction in which ρ is specified.

Theorem 10.7
Any conserved flow f can be decomposed into flows along paths $\gamma_1, \gamma_2, \ldots, \gamma_t$ from S to F and flows around circuits $\rho_1, \rho_2, \ldots, \rho_w$.

Corollary If f and f^* are two conserved flows with the same value ϑ, then f^* may be obtained from f by the addition of flow around cycles $\rho_1, \rho_2, \ldots, \rho_w$ where each ρ_i is flow augmenting with respect to f.

Proof (See exercise 10.4.)

Theorem 10.8
A flow f has minimal cost if and only if there is no flow augmenting cycle ρ with negative length $[l(\rho) < 0]$.

Proof The necessity is obvious.
Suppose now that f^* is a flow with the same value as f but with lower cost. Let $\rho_1, \rho_2, \ldots, \rho_w$ be as in the corollary to theorem 10.7. Since cost $(f^*) = $ cost $(f) + \sum_{i=1}^{w} l(\rho_i) <$ cost (f) it follows that $l(\rho_i) < 0$ for at least one ρ_i. The result follows. □

Theorem 10.9 (Busacker and Gowen, 1961)
If a minimal cost flow f of value ϑ is augmented by ϵ along a minimal length augmenting chain π from S to F then a minimal cost flow f' of value $\vartheta + \epsilon$ is obtained.

Proof Suppose f' is not a minimal cost flow of value $\vartheta + \epsilon$. By theorem 10.8 there must exist a negative length augmenting cycle ρ which has an arc ij in common with π. By taking the arcs, other than ij, from π and ρ a flow augmenting path π' (with respect to f) from S to F can be formed such that $l(\pi') < l(\pi)$. But this contradicts the minimality of π and so the assumption that f' is not a minimal cost flow must be false. □

Theorem 10.8 leads to a method (Klein, 1967) for finding a minimal cost flow of value ϑ from any flow of value ϑ. All that is required is the application of a routine for detecting negative length cycles (chapter 4).

Example 10.8 Find a minimal cost maximal flow through the network of figure 10.9a starting with the initial feasible flow of figure 10.9b.

Solution The flow augmenting cycle $ScbS$ of length -6 may be identified and flow round it increased by 1 unit to give a flow of cost $133 - 6 = 127$ (figure 10.9c). Next the flow augmenting cycle $SbaS$ of length -4 may be identified and flow round it increased by 1 unit to get a flow of cost 123. Finally the flow augmenting cycle $ScaS$ is identified and flow round it increased by 1 unit to get a flow of cost 122 (figure 10.10c). Since no negative length flow augmenting chains remain this latter network flow must be a minimal cost maximal flow.

An alternative method starts with zero initial flow and builds a minimal cost flow of value $\vartheta \leqslant \max_{f} \vartheta_f$ by repeatedly augmenting the flow along minimal cost flow augmenting chains from S to F (see theorem 10.9).

Example 10.9 Find minimal cost flows of values 6, 9 and 12 through the network of figure 10.9a.

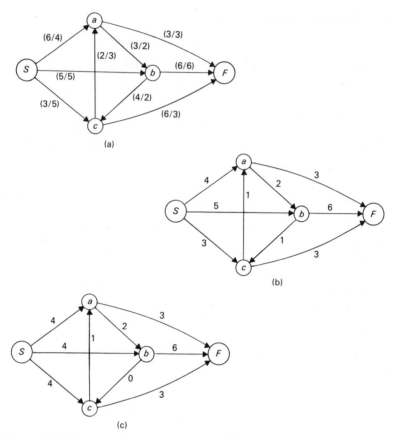

Figure 10.9 Numbers alongside the arcs of (a) are of the form (cost per unit flow/capacity). In (b) and (c) the numbers shown are arc flows

Solution Starting with zero flow in every arc the minimal length flow augmenting chain from S to F is *ScaF* of length 8 and consequently the flow can be augmented by 3 units. Now the shortest flow augmenting chain is *ScF* of length 9 and this permits augmentation of the flow by 2 units (*SaF* has the same length but is no longer flow augmenting).

The next shortest flow augmenting chain is *SacF*, of length $(6-2+6)=10$, and flow along it may be augmented by 1 unit, thereby yielding a minimal cost flow of value 6 (figure 10.10a).

SbF becomes the next flow augmenting chain, and although flow along *SbF* can be augmented by up to 5 units, the augmentation is restricted to 3 units to obtain a minimal cost flow of value 9 (figure 10.10b).

In order to find a minimal cost flow of value 12 (which is in fact a minimal cost maximal flow), flow along *SbF* is augmented by a further 2 units and flow along *SabF* is augmented by 1 unit (figure 10.10c).

Out-of-kilter Method

The use of kilter numbers to obtain a feasible flow in a network with lower bound constraints was shown in section 10.1. A similar approach will now be adopted with a view to finding minimal cost flows.

Let $\{\pi(i)\}$ be a set of numbers, called *vertex* numbers, with one number $\pi(i)$ associated with each vertex i. Then, corresponding to a feasible network flow f, kilter numbers $K_\pi(f_{ij})$ *relative* to $\{\pi(i)\}$ are defined by

$$K_\pi(f_{ij}) = \begin{cases} f_{ij} - l_{ij} & \text{if } \pi(j) - \pi(i) < p_{ij} \\ 0 & \text{if } \pi(j) - \pi(i) = p_{ij} \\ c_{ij} - f_{ij} & \text{if } \pi(j) - \pi(i) > p_{ij} \end{cases}$$

Theorem 10.10
A feasible network flow f of value ϑ is a minimal cost flow of value ϑ if $K_\pi(f_{ij}) = 0$ for all arcs ij.

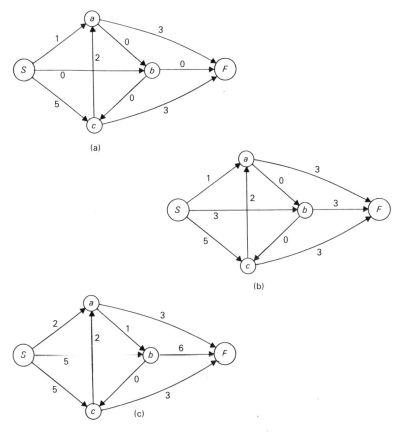

Figure 10.10 Minimal cost network flows of values 6, 9 and 12

Proof Let $\rho \equiv u_0 u_1 u_2 \ldots u_w u_0$ be any flow augmenting cycle. If $u_\alpha u_{\alpha+1}$ is a forward arc of ρ then $c_{u_\alpha u_{\alpha+1}} - f_{u_\alpha u_{\alpha+1}} > 0$ and, since $K_\pi(f_{u_\alpha u_{\alpha+1}}) = 0$, it follows that

$$\pi(u_{\alpha+1}) - \pi(u_\alpha) \leq p_{u_\alpha u_{\alpha+1}} \qquad u_\alpha u_{\alpha+1} \in FA(\pi) \tag{10.12}$$

On the other hand if $u_{\alpha+1} u_\alpha$ is a backward arc of ρ then $f_{u_{\alpha+1} u_\alpha} - l_{u_{\alpha+1} u_\alpha} > 0$ and, since $K_\pi(f_{u_{\alpha+1} u_\alpha}) = 0$

$$\pi(u_\alpha) - \pi(u_{\alpha+1}) \geq p_{u_{\alpha+1} u_\alpha}$$

That is

$$\pi(u_{\alpha+1}) - \pi(u_\alpha) \leq -p_{u_\alpha u_{\alpha+1}} \qquad u_{\alpha+1} u_\alpha \in BA(\pi) \tag{10.13}$$

Summing the inequalities 10.12 and 13 round ρ gives

$$0 = \{ [-\pi(u_0) + \pi(u_1)] + [-\pi(u_1) + \pi(u_2)] + \ldots + [-\pi(u_w) + \pi(u_0)] \}$$

$$\leq \sum_{FA(\rho)} p_{u_\alpha u_{\alpha+1}} - \sum_{BA(\rho)} p_{u_{\alpha+1} u_\alpha} = l(\rho)$$

That is, every flow augmenting cycle has non-negative length and so by theorem 10.8 the flow must be of minimal cost for the specified flow value. □

The reader may find it helpful to think of $\pi(i)$ as being the price at i of 1 unit of the commodity flowing through the network. If then $\pi(j) - \pi(i) > p_{ij}$ it is profitable to ship the commodity from i to j and f_{ij} should be as large as possible; that is $f_{ij} = c_{ij}$ and $K_\pi(f_{ij}) = 0$. Conversely if $\pi(j) - \pi(i) < p_{ij}$ then f_{ij} should be set to l_{ij} again giving $K_\pi(f_{ij}) = 0$. In the remaining case $\pi(j) - \pi(i) = p_{ij}$, $K_\pi(f_{ij}) = 0$ anyway and f_{ij} is only constrained to lie between l_{ij} and c_{ij}.

The following algorithm adjusts the flows so as to reduce the kilter numbers relative to a particular set of vertex numbers. When no further reduction is possible, but not all kilter numbers are zero, then a more promising set of vertex numbers is derived. The process is repeated until no arc is out of kilter and hence a minimal cost flow has been found.

Algorithm (OOK)
(To find a minimal cost flow of value ϑ)

Step 1 (Setup)
Let $f = \{ f_{ij} \}$ be some feasible flow of the desired value, and $\{ \pi(i) \}$ any set of vertex numbers.

Step 2 (Termination)
Select an out-of-kilter arc ab, if there are any; otherwise terminate.

Network Flow: Extensions

Step 3 (Labelling)
 Label b with $(+a)$ if $\pi(b) - \pi(a) > p_{ab}$;
 else label a with $(-b)$.
 Repeatedly perform step 3a while labelling is possible and either a or b is not labelled.

 Step 3a
  j is labelled with
   $(+i)$ if i is labelled, $\pi(j) - \pi(i) \geqslant p_{ij}$ and $f_{ij} < c_{ij}$;
   $(-k)$ if k is labelled, $\pi(k) - \pi(j) \leqslant p_{jk}$ and $f_{jk} > 0$.

  (1) If a and b are both labelled go to step 4;
  (2) If no further labelling possible go to step 5.

Step 4 (Update flows)
 Augment the flow (as much as possible) round the cycle which includes a and b as determined by the labels.
 Go to step 2.

Step 5 (Update vertex numbers)
 Increase $\pi(i)$ by δ for each unlabelled i.
 Go to step 2.

The choice of the value of δ in step 5 is discussed below.

Example 10.10 Consider again the problem of finding a minimal cost maximal flow starting with the flow of figure 10.9b.

Solution Choose a set of vertex numbers, say $\pi(S) = 0$, $\pi(a) = 6$, $\pi(b) = 6$, $\pi(c) = 6$, $\pi(F) = 12$. The kilter numbers are then as shown in figure 10.11a.

Step 2 Select out-of-kilter arc ab.
 3 Assign label $(-b)$ to a, label $(-a)$ to c, label $(-a)$ to S and label $(-c)$ to b.
 4 Augment the flow round $bacb$ by 1 unit.
 2 ab is still out-of-kilter and so we reselect it.
 3 Assign label $(-b)$ to a, label $(-a)$ to S and label $(+S)$ to c. No more labelling possible.
 5 δ is to be added to each unlabelled vertex.
  Taking $\delta = 3$, allows another vertex (b) to be labelled. Adjust vertex numbers and return to step 2 (figure 10.11b).
 2 Only arc Sc is out of kilter.
 3 Assign label $(+S)$ to c.
  No more labelling possible.

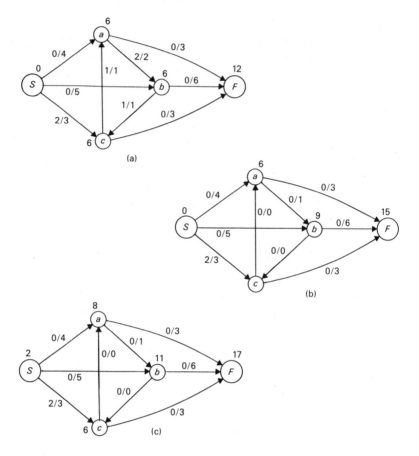

Figure 10.11 Numbers alongside arcs are of the form (kilter number)/flow. Vertex numbers π_i are given beside each vertex

5   δ is to be added to each unlabelled vertex.
Taking $\delta = 2$ allows another vertex (a) to be labelled. Adjust vertex numbers and return to step 2 (figure 10.11c).
2   Sc is still out of kilter — reselect.
3   Assign label (+S) to c, label (+c) to a, label (+a) to b and label (−a) to S.
4   Augment flow round $ScaS$ by 2 units.
2   No out-of-kilter arcs remain. The flow has minimal cost.

The choice of the initial set of vertex numbers is arbitrary and $\pi(i) = 0$ for all i will do. However, the algorithm will work more efficiently if vertex numbers are chosen so that the kilter numbers are small (exercise 10.5).

Network Flow: Extensions 237

It may be noticed from the above example that δ is chosen 'as small as possible' but so as to allow one more vertex to be labelled. (Other vertices may of course be labelled as a consequence of this.) In order to formalise this two sets A_1 and A_2 are defined

$$A_1 = \{ \; ij \; | \; i \text{ labelled}, j \text{ unlabelled and } \pi(j) - \pi(i) < p_{ij} \}$$
$$A_2 = \{ \; ij \; | \; i \text{ unlabelled}, j \text{ labelled and } \pi(j) - \pi(i) > p_{ij} \}$$

Let $\quad \delta_1 = \min_{ij \in A_1} (p_{ij} - [\pi(j) - \pi(i)])$

$\quad \delta_2 = \min_{ij \in A_2} ([\pi(j) - \pi(i)] - p_{ij})$

where $\delta_i = \infty, i = 1, 2$ if A_i is empty. Then the value chosen for δ is the smaller of δ_1 and δ_2

$$\delta = \min(\delta_1, \delta_2)$$

With this choice of δ the algorithm will terminate in a finite number of steps.

The requirement that the initial flow be feasible is often not insisted upon, and an out-of-kilter algorithm which is essentially a combination of the out-of-kilter algorithm of section 10.1 (e) for finding feasible flows and algorithm (OOK), is used. For further details of this and other refinements the reader is referred to Lawler (1976) and Barr *et al*. (1974). These latter authors describe an implementation of their ideas as the FORTRAN program SUPERK which they found to be faster in practice than other codes available at that time. This was particularly so as problem size and/or the arc density [that is, (number of arcs)/(possible number of arcs)] increased. Some times obtained using SUPERK, on a CDC 6600 computer, are given in section 10.3 in table 10.1.

Non-linear Minimal Cost Flow Problems

The assumption that the cost functions p_{ij} are linear functions of the corresponding flows f_{ij} may well be a good approximation for many problems; for example, if the arcs can be well represented as a bundle of separate lines each of which can carry 1 unit of flow. However, as motor vehicle drivers know, in road transport the time and hence the cost of a journey increases as the volume of traffic increases. Consequently the cost functions will be convex functions of arc flows; the form adopted for U.S. Federal Highway Administration traffic assignment models is

$$p_{ij}(f_{ij}) = a_{ij} + b_{ij}(f_{ij})^4$$

Not surprisingly a different approach is required to find a minimal cost flow of a particular value ϑ (or for a set of values ϑ^α in the multi-commodity case).

Suppose that a flow $\tilde{f} = \{ \tilde{f}_{ij} \}$ is known and that the total cost is $C(\tilde{f})$.

Then approximate $C(f)$ by a linear function

$$C(f) \approx C(\tilde{f}) + \sum_{ij} \frac{\partial C(\tilde{f})}{\partial f_{ij}}(f_{ij} - \tilde{f}_{ij})$$

$$= C(\tilde{f}) + \nabla C(\tilde{f}) \cdot (f - \tilde{f})$$

Since $C(\tilde{f}) - \nabla C(\tilde{f}) \cdot \tilde{f}$ is constant the following linear programming problem would yield a minimal cost solution if the cost function were indeed linear

(P1)   minimise $\nabla C(\tilde{f}) \cdot f$

subject to

$$\sum_{i \in \Gamma^{-1}_j} f_{ij} - \sum_{k \in \Gamma j} f_{jk} = \begin{cases} -\vartheta & j = S \\ 0 & j \neq S, F \\ \vartheta & j = F \end{cases}$$

$$0 \leq f_{ij} \leq c_{ij}$$

This is the problem, slightly disguised, whose solution was considered above. However, since $C(\tilde{f})$ is not linear the solution to this LP problem will not be a minimal cost solution, but we might expect it to yield a good direction [that of $(f^* - \tilde{f})$] in which to search for a better solution than $\tilde{f}$. The next step is to perform a 1-dimensional search in the chosen direction $(f^* - \tilde{f})$ to find a minimum solution (to within a specified tolerance) to

(P2) $\underset{0 \leq \theta \leq 1}{\text{minimise}}$ $C[\tilde{f} + \theta(f^* - \tilde{f})]$

The procedure can now be repeated with the improved solution $\tilde{f} + \theta(f^* - \tilde{f})$ replacing $\tilde{f}$. This algorithm is a special case of a more general method due to Frank and Wolfe (Zangwill, 1969) which is applicable to problems with a non-linear convex objective function but linear constraints.

Algorithm (FW)
(To find a minimal cost flow: non-linear case)

Step 1 (Setup)
   Let $\tilde{f}$ be a feasible flow. Specify tolerances ϵ_1 and ϵ_2.

Step 2 (Iteration)
   Find a solution f^* to problem (P1).
   Find a value of θ, optimal to within ϵ_2 for problem (P2).

Step 3 (Termination)   If $\theta = 0$ terminate.
   If $\delta C = |\{C[\tilde{f} + \theta(f^* - \tilde{f})] - C(\tilde{f})\}| < \epsilon_1$ terminate;
   else set $\tilde{f} \leftarrow \tilde{f} + \theta(f^* - \tilde{f})$ and go to step 2.

Example 10.11 Find a minimal cost flow ($\epsilon_1 = 0.01$) of value 5 through the network of figure 10.12. For illustration the simpler cost functions $p_{ij} = a_{ij} + b_{ij}(f_{ij})$ will be used.

Solution In order to simplify the presentation, arcs will be given by a single number as shown in figure 10.12. At zero flow the cheapest route by far is along arcs e_5 and e_6. We will take as a (not very good) initial solution

$$\tilde{f} = (\tilde{f}_1, \tilde{f}_2, \ldots, \tilde{f}_7) = (0,5,0,0,0,5,0)$$

$$C(\tilde{f}) = [(1)(5) + (0.3)(25)] + [(2)(5) + (0.3)(25)] = 30$$

$$\nabla C(\tilde{f}) = [\frac{\partial C(\tilde{f})}{\partial f_1}, \frac{\partial C(\tilde{f})}{\partial f_2}, \ldots, \frac{\partial C(\tilde{f})}{\partial f_7}]$$

$$= [a_1 + 2b_1(f_1), \ldots, a_7 + 2b_7(f_7)] = (2,4,1,1,1,5,2)$$

Using these components of ∇C as the new unit costs, problem (P1) is solved. The cheapest route is via e_1, e_4 and e_7 giving

$$f^* = (f_1^*, f_2^*, \ldots, f_7^*) = (5,0,0,5,0,0,5)$$

$$C[\tilde{f} + \theta(f^* - \tilde{f})] = 30 - 20\theta + 32.5\theta^2$$

The minimum of this function occurs at $\theta = 0.31$. Substituting gives

$$\tilde{f} \leftarrow (1.55, 3.45, 0, 1.55, 0, 3.45, 1.55), \quad C(\tilde{f}) \leftarrow 26.9$$

This completes the first iteration. The next two iterations yield

$\theta = 0.204, \tilde{f} \leftarrow (1.23, 3.77, 0, 1.23, 1.02, 2.75, 2.25), C(\tilde{f}) \leftarrow 26.53$

$\theta = 0.0033, \tilde{f} \leftarrow (1.226, 3.774, 0, 1.226, 1.017, 2.757, 2.243), C(\tilde{f}) \leftarrow 26.531$

If the actual arc flow values are of interest then perhaps a more appropriate

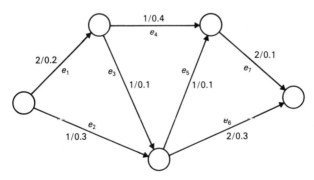

Figure 10.12 Numbers alongside each arc e_j are of the form a_j/b_j (see example 10.11)

alternative termination criterion is that the maximum change in value of any arc flow should be less than α per cent of the total flow value (LeBlanc et al., 1975).

Finally it might be noted that the above method extends in a straightforward way to multi-commodity flows (LeBlanc et al., 1975, Daganzo, 1977a,b).

10.3 THE SIMPLEX METHOD APPLIED TO NETWORK PROBLEMS

Somewhat more knowledge of linear programming will be assumed in this section. A standard text that is particularly useful in the present context is Dantzig (1963).

The Transshipment Problem

A transportation problem was introduced in example 10.1 in which was sought the most economical way of directly supplying the specified amount of goods from the supply points to the demand points. However, situations arise in which it may be profitable to allow the commodity to pass through other sources and destinations and indeed points where there is no net supply or demand. This gives rise to the transshipment problem (TRP) which may be formulated

TRP minimise $\sum_{i} \sum_{j \in \Gamma i} p_{ij} f_{ij}$ (10.14a)

subject to

$$\sum_{i \in \Gamma^{-1}j} f_{ij} - \sum_{k \in \Gamma j} f_{jk} = a_j \quad j \in X - \{t\}$$ (10.14b)

$$f_{ij} \geq 0, \ j \in \Gamma i$$ (10.14c)

where

(X, Γ) is the underlying transport network
f_{ij} is the flow in arc ij
p_{ij} is the cost of sending one unit of flow along ij
a_j is the net requirement at j (that is, demand minus supply)
t is any specified vertex. (Flow is conserved at t of course but this follows from the remaining conservation equations.)

By adding a super source and super sink (section 10.1) this could be converted to a standard form minimal cost flow problem which could be solved by the methods of section 10.2. Alternatively, TRP could first be converted to a transportation problem and solved by the 'stepping stones' method (Taha, 1976). In this section we shall look at the application of the (primal) simplex method to the LP formulation above, and how this might be achieved efficiently by making use of the structure of the problem. First the special nature of a simplex basis for the problem is examined. Let flow f be a basic solution for

TRP. Then if B is the set of basic variables let G_B be the corresponding set of arcs

$$G_B = \{ ij \mid f_{ij} \in B \}$$

Theorem 10.11
G_B forms a spanning tree of the graph (X, Γ) (the directions of the arcs being ignored).

Proof If (X, Γ) has n vertices G_B will contain $n-1$ arcs, one for each of the equations of the independent set 10.14b. Suppose now that a cycle $\sigma = u_1 u_2 \ldots u_{r+1}$ $(u_1 = u_{r+1})$ can be formed where $u_\alpha u_{\alpha+1}$ or $u_{\alpha+1} u_\alpha$ belongs to G_B. Assume that the flow in each arc of σ is non-zero. (Readers are left to consider the case of a degenerate basis.) Two new flows f^+ and f^- may be obtained by sending a small additional flow of value $\epsilon > 0$ around σ in the 'forward' and 'reverse' directions respectively. Now $f^+ \neq f \neq f^-$ but $f = \frac{1}{2}(f^+ + f^-)$ indicating that f is a proper linear combination of the distinct feasible solutions f^+ and f^-. This contradicts the assumption that f is basic and so the assumption that a cycle σ exists must be false.

Since G_B contains $n-1$ elements and no cycles it must be a tree which clearly spans (X, Γ). □

From 10.14 the dual formulation of TRP is

$$\text{maximise} \quad \sum_{j \in X} a_j \pi_j \qquad (10.15a)$$

subject to

$$\pi_j - \pi_i \leq p_{ij} \qquad (10.15b)$$

$$\pi_i \text{ unrestricted in sign} \qquad (10.15c)$$

and at optimality, primal and dual variables must satisfy the complementary slackness conditions

$$\pi_j - \pi_i = p_{ij} \quad \text{if } f_{ij} > 0 \qquad (10.16a)$$

$$f_{ij} = 0 \quad \text{if } \pi_j - \pi_i < p_{ij} \qquad (10.16b)$$

The solution strategy is to start with a primal feasible solution f corresponding to basis B then to determine values of dual variables π_i (or vertex numbers in the terminology of section 10.2). If π_t is set arbitrarily (to 0 say) then the values of the vertex numbers are determined uniquely (since the arcs of G_B form a tree). Optimality is checked against conditions 10.16. If an out-of-kilter arc ab (for which $\pi_b - \pi_a > p_{ab}$) exists then this arc is brought into the basis. (Strictly it is the corresponding flow f_{ab} which comes into the basis, but it is convenient to extend to elements of G_B the use of such terms as 'basic', 'non-basic', 'enters the basis', etc.) Now $G_B \cup \{ab\}$ contains a unique cycle, σ, and by sending a flow ϵ around σ in the direction of ab some arc rs will become flowless provided ϵ is large enough. This is the arc that leaves the basis, and a tree is restored.

Example 10.12 Solve the TRP of figure 10.13a in which the number alongside arc ij is p_{ij} and the requirement a_j is shown by each vertex j.

Solution Suppose that we start with the basic solution illustrated by the solid arcs of the spanning tree shown in figure 10.13b. For this solution the non-zero arc flows are

$$f_{23} = 5, f_{41} = 8, f_{52} = 4, f_{54} = 7, f_{65} = 10$$

giving a value of 195 for the objective function. Forming $p_{ij} - (\pi_j - \pi_i)$ it is found that f_{43} is a candidate to enter the basis. Another possibility is f_{61} (see exercise 10.8). Which variable should leave? Addition of arc 43 leads to the formation of cycle $\sigma = 43254$ and so either f_{54}, f_{52} or f_{23} becomes non-basic. Suppose a flow of value ϵ is sent round σ. The flow pattern remains feasible as ϵ is increased up to the value 4 at which point arc 52 becomes flowless. Correspondingly f_{52} is the leaving variable and the new basic variables and their values are

$$f_{23} = 1, f_{41} = 8, f_{43} = 4, f_{54} = 11, f_{65} = 10$$

The length of σ is $4 - 9 - 3 + 4 = -4$ and so the value of the new basic solution is $195 + 4 \times (-4) = 179$ as may be verified directly.

f_{61} is now the entering variable and f_{41} the leaving variable. The flow of 8 is sent round 61456 giving a new basic solution (figure 10.13d) which may be verified as being optimal and with value 115.

The reader may see how the above approach resolves to the familiar stepping stones method in the special case of a transportation problem.

The above example shows how the simplex method may be applied to TRP by considering only the basis trees G_B together with the original arc costs. We now look more closely at how this might be implemented efficiently.

First, it may be noted that in order to perform a basis change it is required to traverse a cycle consisting of arcs which may be oriented 'with the cycle or in the opposite direction'. In order to facilitate this process it is convenient to store bases in the form of *rooted precedence trees*. For these, one vertex, u say, is selected as the *root*, then the tree built up recursively using the following algorithm.

Algorithm (PT)
(To form precedence trees)

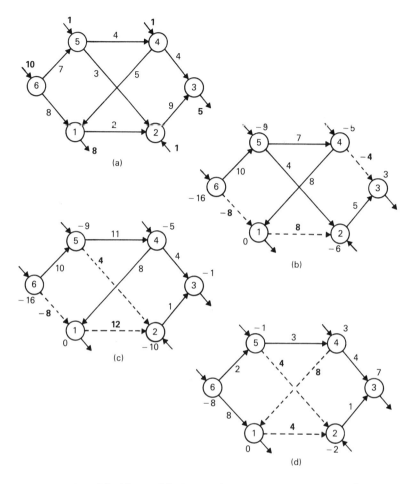

Figure 10.13 For (b), (c) and (d) the number alongside arc ij is the flow f_{ij} if it is a basic arc (full line) and $p_{ij} - (\pi_j - \pi_i)$ if it is a non-basic arc (broken line). Vertex numbers are shown beside each vertex in (b), (c) and (d)

Step 1 (Setup)
  Set $V = \{u\}$, $A = \phi$, depth $(u) = 0$, father $(u) = 0$.
  Repeat step 2 until $|A| = n - 1$.

Step 2 (Iteration)
  For each arc $ij \in B$ such that $i \in V$, $j \notin V$
  set $V \leftarrow V \cup \{j\}$, $A \leftarrow A \cup \{ji\}$, depth (j) = depth $(i) + 1$, father $(j) = i$.
  For each arc $ij \in B$ with $i \notin V$ and $j \in V$
  set $V \leftarrow V \cup \{i\}$, $A \leftarrow A \cup \{ij\}$, depth (i) = depth $(j) + 1$, father $(i) = j$.

Step 3 (Termination)

(V, A) is the required precedence tree corresponding to basis B and will be denoted by PG_B.

It should be noted that all arcs of the precedence tree 'point towards the root u', and some will have a direction opposite to that of the corresponding arcs of the original network.

To detect the cycle that results when an arc, ab say, is added to a basic set of arcs is quite straightforward. We assume that depth $(a) \geq$ depth (b) (if this is not so a similar argument obtains). Trace a path back from a via the precedence (father) pointers for depth (a) − depth (b) arcs to a vertex a'. Now a' may coincide with b in which case σ is found [the path from a to a' ($=b$) together with ab]. If this is not the case check whether father (a') = father (b); if not check whether father [father (a')] = father [father (b)] and so on until equality is obtained (as of course it must be, even if only at the root vertex).

By checking on the flow in arcs which will be oriented in the opposite direction to ab in σ, the amount by which the flow round σ is to be changed can be determined at the same time that σ is being found.

Suppose ab is the incoming arc and rs the outgoing arc. Then it is easily seen that, in PG_B, either

(i) the path from a to the root vertex contains arc rs; or
(ii) the path from b to the root contains arc sr.

Denote by P^* the path from a to r and b to s in cases (i) and (ii) respectively. In order to update the precedence tree PG_B on change of basis

(1) reverse the arcs of P^*;
(2) set father $(a) = b$ if case (i) holds
    father $(b) = a$ if case (ii) holds.

But what of updating the vertex numbers? The new basis can be regarded as two subtrees T_1, T_2 connected by arc ab. It is easily seen that if vertex numbers are calculated again from scratch then those attached to the subtree which contains the root (T_1 say) will be as before. Hence updating may be restricted to the vertices of T_2 starting at b. In fact, as observed by Srinivasan and Thompson (1972), the vertex numbers could be left as they are for T_2 and those for T_1 updated working from a. They went a step further and, by storing certain extra information, always updated the vertex numbers for the smaller of the two trees.

An alternative system described by Glover *et al.* (1972) is the *augmented predecessor index method*. With this, extra pointers are kept which permit easy tracing of paths and subtrees in the opposite direction to that specified by the precedence pointers. However, these new pointers are not just the reverse of the old pointers but are based on Johnson's (1966) triple labelling scheme. This is perhaps best illustrated in terms of family relationships. Thus if j = father (i)

then j is the 'father' of i and each such i is a 'child' of j. A pointer eldest (j) is provided from j to one of its children [that is, eldest $(j) = i_1$ for some i_1 such that father $(i_1) = j$]. Also there is a pointer brother (i_1) from i_1 to the second eldest child and so on down to the youngest child i_r say. The ends of all lists are indicated by a pointer to some special object [for example, brother $(i_r) = 0$]. An example, in which the end of list pointers are not shown, is given in figure 10.14c for the graph PG_B of figure 10.13b. This triple labelling system is easy to implement using a FORTRAN-like computer language, with three length n arrays *father, eldest* and *brother*. Such a system allows very efficient traversal of the subtree at b and hence very efficient updating of vertex numbers. If ab is the incoming arc and $\delta = p_{ab} - (\pi_b - \pi_a)$ then vertex numbers π_i are updated by setting

$\pi_i \leftarrow \pi_i + \delta$ if i is in new subtree at b
$\pi_i \leftarrow \pi_i$ otherwise

On the other hand, restructuring the tree after a basis change, though quite efficient, is somewhat complicated (Glover *et al.*, 1972).

These ideas have been incorporated in the computer program PNET developed by Glover and his collaborators at the University of Texas. A further refinement (Glover *et al.*, 1974) is to replace the augmented predecessor index method by using 'threaded' precedence trees. Roughly speaking the thread of

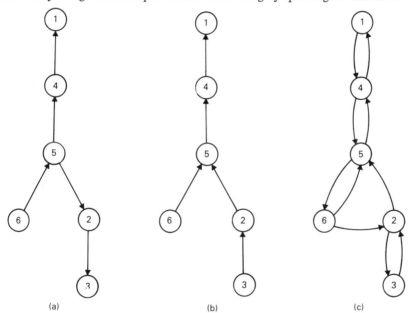

Figure 10.14 The graph G_B for the basis of figure 10.13b is shown in (a) and the corresponding graph PG_B is shown in (b) if vertex 1 is selected as the root. (c) illustrates the augmented predecessor index structure with father, son and brother pointers (end of list pointers are not shown)

a vertex in a precedence tree is the next new vertex to be encountered during a depth-first search of that tree. [For example, with reference to figure 3.4b, thread $(L) = b$, thread $(b) = c, \ldots,$ thread $(h) = q$, thread $(q) = g, \ldots$.] Traversing subtrees is thus made extremely efficient. The restructuring of the tree is also efficient but complex to state and the reader is referred to Glover *et al.* (1974) for details.

A computer program, called PNET-1, was developed from PNET using the *augmented threaded list* structure instead of the augmented predecessor index method. Glover *et al.* (1974) give results for the two programs, and a selection of these results is given in table 10.1 together with comparative results using SUPERK (see section 10.2). Times given are all in seconds on a CDC 6600 computer. For further details of the problems see Klingman *et al.* (1974).

Table 10.1

Problem number	Vertices	Arcs	PNET	PNET-1	SUPERK
1	100 + 100	1300	1.30	1.17	5.68
5	100 + 100	2900	1.88	1.73	6.77
6	150 + 150	3150	3.55	3.06	11.05
10	150 + 150	6300	5.88	5.57	14.13
16	400	1306	2.40	2.15	5.27
27	400	2676	4.42	4.31	7.50
28	1000	2900	6.35	5.67	13.91
31	1000	4800	9.59	8.48	17.05
32	1500	4342	15.70	13.59	22.88
35	1500	5730	19.39	17.13	29.96

If arcs are capacitated, the above development goes through since, for bounded variable problems, variables may be non-basic with value zero or non-basic at their upper bound (Taha, 1976).

For further reading on implementing transshipment algorithms see Bradley *et al.* (1977).

Generalised Networks

We now turn to so-called generalised networks. These are in effect transshipment networks with gains on arcs permitted (section 10.1) and have the form (Glover *et al.*, 1978)

$$\text{minimise} \sum_i \sum_{j \in \Gamma_i} p_{ij} f_{ij} \tag{10.17a}$$

subject to

$$\sum_{i \in \Gamma^{-1} j} g_{ij} f_{ij} - \sum_{k \in \Gamma j} f_{jk} = a_j \qquad (10.17b)$$

$$0 \leq f_{ij} \leq c_{ij} \qquad (10.17c)$$

The case $i = j$ is allowed with $g_{ii} = 0$ and corresponds to a loop.

Glover *et al.* (1978) assert that practical settings in which generalised network problems arise include problems of resource allocation, production, distribution, scheduling, capital budgeting and so on.

Definition 10.1
A *quasi-tree* is a connected graph on n vertices with precisely n edges (or arcs) — it will contain precisely one cycle.

Theorem 10.12
Each component of the basis graph G_B corresponding to basis B for a generalised network problem is (in general) a quasi-tree.

Proof Consider any component C of G_B. The variables associated with the arcs of C appear only in equation 10.17b corresponding to vertices $j \in C$. These equations could not, in general, be satisfied if there are fewer variables than equations. Since this is true for each component the total number of arcs in G_B is at least as great as the number of equations. However, the number of elements in B is equal to the number of equations. Hence the number of arcs in each component C is equal to the number of vertices in that component. It follows that each component of G_B must be a quasi-tree. □

How then can the data structures used for the 'pure' network problems of the previous section be adapted to the more general situation? First, for each quasi-tree C and each vertex j of C not on the cycle, father (j) is set as if the cycle were condensed to a single vertex to form the root of a precedence tree. Finally the pointers father (j) for j on the cycle are set all 'clockwise' or all 'anticlockwise' thus forming a circuit. This gives the *precedence graph PG_B*. For each $i \in C$ the set $\{ j \mid \text{father}(j) = i \}$ are brothers one of whom is regarded as the eldest and designated eldest (i).

Other points to be resolved are: which arcs will have their flows changed if arc ab is the arc incoming to the present basis? How is a basis graph G_B updated? How are vertex numbers updated?

Example 10.13 The precedence graph PG_B for a generalised network problem is given in figure 10.15. If now the basis is to be changed with f_{17} as the incoming variable which will be the outgoing variable, given the information below which relates to the current basis?

arc ij	18	63	65	59	69	27	25	17	83
g_{ij}	½	1	4	¼	2	⅓	½	2	1
f_{ij}^E	9	–	2	–	5	38	–	–	3
w_{ij}^E/ϵ	–1	½	–¼	2	–¼	–6	6	1	–½

Solution We start by assuming that the entering flow w_{17}^E in arc 17 will be raised from 0 to ϵ when arc 17 becomes basic. To balance this requirement of ϵ at 1, an entering flow $w_{18}^E = -\epsilon$ must be assigned to arc 18 (that is, the entering flow must be reduced by ϵ). Hence $w_{18}^L = -g_{18}\epsilon$ and so to balance flow at vertex 8, $w_{83}^E = -g_{18}\epsilon$ and $w_{83}^L = -g_{18}g_{83}\epsilon$. To balance flow at vertex 3 we must have $w_{63}^L = -w_{83}^L = g_{18}g_{83}\epsilon$ and $w_{63}^E = g_{18}g_{83}\epsilon/g_{63}$. Then from this $w_{65}^E = -w_{63}^E$ and $w_{65}^L = (-g_{18}g_{83}g_{65}\epsilon)/g_{63} = -2\epsilon$. On the other hand a flow of $w_{17}^E = \epsilon$ in arc 17 generates a leaving flow of $w_{17}^L = g_{17}\epsilon$ and leads on to a flow of $w_{25}^L = g_{17}g_{25}\epsilon/g_{27} = 3\epsilon$ to vertex 5. Thus a net flow of $3\epsilon - 2\epsilon = \epsilon$ out of vertex 5 is required. This 'excess' flow can only be absorbed by the flow round the circuit. It is readily verified that a flow of θ leaving 5 along 59 becomes $g_{59}g_{65}\theta/g_{69}$ on returning to 5 along arc 65. Thus in order to absorb an amount of flow ϵ it is necessary to set

$$\theta = \epsilon/(1 - g_{59}g_{65}/g_{69}) = 2\epsilon$$

The term multiplying ϵ in the above expression is called the *loop factor*.

The flows w_{ij}^E resulting from setting $w_{17}^E = \epsilon$ are as shown in the table above.

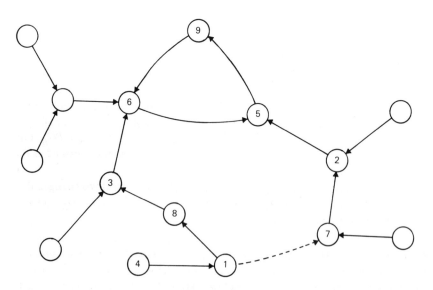

Figure 10.15

Now the value of ϵ is increased until the flow $f_{ij}^E + w_{ij}^E(\epsilon)$ becomes zero for (at least) one arc which will correspond to the variable leaving the basis. For the above example it is seen that arc 83 is the first arc to become flowless when $\epsilon = 6$. The new precedence tree is then formed by setting

father(8) = 1, father(1) = 7,

that is adding arc 17, deleting arc 83 and replacing arc 18 by arc 81.

Vertex numbers $\{\pi_i\}$ may be recomputed as soon as it is known that arc 83 corresponds to the outgoing variable for they satisfy $g_{ij}\pi_j - \pi_i = p_{ij}$ for basic arcs and all but π_1, π_8, π_4 remain as they are. These three numbers are readily obtained from the new precedence graph and the old value of π_7.

Returning now to the general situation we note that the incoming arc may connect two vertices which may be on the same or different components of PG_B and

(1) neither lies on a circuit (as above); or
(2) one lies on a circuit; or
(3) both lie on circuits.

Moreover, there are a variety of ways in which the leaving arc may disconnect a circuit. However, despite all this apparent complexity, precedence graphs can be updated by simple rules. Let i be any vertex then the backward path P_i from i (in PG_B) is obtained by tracing back via father(i), father [father(i)], etc., as far as possible without repeating an arc. That is, the formation of P_i stops as soon as a vertex is encountered for the second time. Next if ab is the incoming arc then a flow of 1 ($\epsilon = 1$) is temporarily assigned to ab giving rise to requirements of 1 at a and $-g_{ab}$ at b. These requirements are transmitted along the backward paths P_a and P_b. If P_a and P_b intersect at vertex t then the net requirement is propagated along the remainder of the backward paths from t.

In example 10.13, 17 is the incoming arc, $P_1 = 1836596, P_7 = 725965$ and $t = 5$. A requirement of 1 at vertex 1 transmits to a requirement of 2 at 5 and a requirement of -2 $(= -g_{17})$ at 7 transmits to a requirement of -3 at 5. The net requirement of -1 at 5 is transmitted round the circuit to give a requirement of ½ at 5 giving a loop factor of $1/(1 - ½) = 2$. The resulting flows w_{ij}^E in $P_a \cup P_b$ are compared with the existing flows f_{ij}^E to determine the value of ϵ which makes the flow $f_{ij}^E + \epsilon w_{ij}^E$ become zero in (at least) one arc. From this the flows can be updated and the leaving arc determined. It might be noted here that in the special case where all gains are 1 then the net requirement at t will always be zero (compare the stepping stones loop of transportation problems).

The precedence graph may be updated in a way analogous to that described for TRP (p. 244). As far as updating vertex numbers is concerned it should be noted that this may be restricted to those vertices i for which a path from i to r (or s) exists in the new tree. Discussions on how this might be achieved are given in Glover *et al.* (1973) and Glover and Klingman (1973).

Glover et al. (1978) report a computer program NETG to solve generalised network problems. They gave a comparison of NETG and APEX-III (a sophisticated general linear programming code maintained by CDC) on 7 problems ranging in size from 100 vertices and 1000 arcs to 1000 vertices and 6000 arcs. Times were not given, but instead the comparison was made in terms of dollar costs, reflecting an amalgam of CPU time, input/output operations performed and central memory used. The results indicated that NETG was from 6 to over 50 times as fast on these problems (in 2 cases APEX-III was terminated prematurely).

Linear Assignment Problem

AP has been introduced in various settings in section 3.1, section 7.1 and chapter 7 exercises. It may also be regarded as a special case of the transportation problem with a set, S, of n sources and a set, D, of n demand points. The requirement at each demand point is 1 and the supply at each source is also 1. The AP may be formulated as the LP

$$\text{minimise} \sum_{\substack{i \in S \\ j \in D}} c_{ij} x_{ij}$$

subject to

$$\sum_{i \in S} x_{ij} = 1, \quad j \in D$$

$$\sum_{j \in D} x_{ij} = 1, \quad i \in S$$

$$x_{ij} \geq 0, \quad i \in S, \ j \in D$$

Although AP can be solved as a transportation problem it should be noted that any basis will have $2n-1$ elements whereas the number of non-zero elements in a basic feasible assignment can be seen to be n. Thus $n-1$ basic variables must be zero. This high degree of degeneracy can be time consuming when using a primal simplex method as several bases corresponding to the same extreme point of the feasible region may be examined before moving on to an adjacent extreme point. Barr et al. (1977) showed how this difficulty can be reduced considerably.

Let f_{ab} be the incoming variable with $a \in S$, $b \in D$ and $a = u_1 u_2 \ldots u_{2p} = b$ be the unique elementary chain in G_B from a to b. Then $u_1, u_3, \ldots, u_{2p-1} \in S$ and $u_2, u_4, \ldots, u_{2p} \in D$.

Definition 10.2
A basis B is *alternating* if it has the properties

(1) the root of G_B is in S;
(2) if arc rs is in G_B and $f_{rs} = 0$ then $r \in S$, $s \in D$;
(3) if arc rs is in G_B and $f_{rs} = 1$ then $r \in D$, $s \in S$.

Now if the basis is alternating and ab is the incoming arc then it is readily verified that

$$f_{u_1 u_2} = f_{u_3 u_4} = \ldots = f_{u_{2p-1} u_{2p}} = 1$$

$$f_{u_3 u_2} = f_{u_5 u_4} = \ldots = f_{u_{2p-1} u_{2p-2}} = 0$$

and $f_{u_1 u_{2p}} = 0$ since it is non-basic. It may be checked that if the arc au_1 is chosen as the outgoing arc then the new basis is also alternating. Moreover the flow values do not change unless b is a descendant of a in the precedence tree in which case flows of 0 and 1 are interchanged round the cycle.

Making basis changes in the way indicated leads to a series of alternating bases. Barr et al. (1977) showed that this sequence must lead to an optimal solution being obtained and that cycling cannot occur. Thus the algorithm (AB) below which they put forward is seen to be valid if it is noted that a first feasible alternating basis can always be found.

Algorithm (AB)
(To solve assignment problems)

Step 1 (Setup)
Select a feasible alternating basis.
Determine vertex numbers.

Step 2 (Termination)
Select a non-basic variable f_{ab} with $p_{ab} - (\pi_b - \pi_a) < 0$.
If there are none such then terminate with an optimal solution.

Step 3 (Iteration)
Determine the unique elementary chain in the precedence tree between a and b.
Let u_1 be, as defined earlier, the vertex adjacent to a on this chain.
Bring f_{ab} into the basis and drop f_{au_1}.
Update PG_B and vertex numbers accordingly.

Since flows are determined uniquely by the depth of the corresponding arc in PG_B, Barr et al. did not store flow values at all. Moreover all arcs rs in which r = father (s) were condensed to a point without losing essential information while reducing storage required and the time spent searching PG_B. They implemented the algorithm, using precedence (father) and thread pointers and the depth function, in FORTRAN on a CDC 6600 computer. Five 200 x 200 assignment problems, with the number of arcs ranging from 1500 to 4500, were solved in a total time of 6.86 s processing time. The code was not optimised for the computer used as was its nearest rival. Taking this and other factors into consideration Barr et al. concluded that '..., it would appear by conservative estimate that the AB algorithm is likely twice as fast as other algorithms for solving assignment problems'.

EXERCISES

10.1 A telephone exchange, in order to maintain a specified level of service (say no more than α per cent of callers are delayed by more than β minutes) has a minimum requirement of s_i operators in the i th ½ hour period of the day $i = 1, 2, \ldots, 48$. There is a list of m shifts (not all with the same duration and which start at various times of the day). Let c_j be the unit cost for shift j and x_j the number of operators assigned to that shift. The problem is to minimise the total cost while maintaining the specified level of service. Formulate this as a minimal cost flow problem in a network N with vertex set $\{1, 2, \ldots, 48\}$ and arc set

$\{ ij \mid i=1, \ldots, 48 \ \& \ j=i+1 \} \cup \{ mk \mid$ a shift covering periods k to m exists $\}$

Segal (1974) shows how a variety of other constraints may be incorporated and also extends the model to take account of relief periods for operators.

10.2 Prove the results of theorems 10.3 and 10.6.

10.3 Three supply points S_1, S_2 and S_3 can supply goods, at rates up to 13, 11 and 4 units per week respectively, to demand points F_1, F_2 and F_3 through the communication network shown in figure 10.16. The number alongside arc xy is the capacity c_{xy}; there is a limitation of a maximum flow of 13 through A and of 8 through B.

An initial flow is obtained by sending 8 units from S_1 to A to F_1, 6 units from S_2 to B to F_2 and 4 units from S_3 to D to F_3. Using this initial flow and the Ford–Fulkerson algorithm find a maximal network flow.

Verify the maximality of your flow by finding a minimal cut.

(LU 1977, abridged)

10.4 Prove the result of the corollary to theorem 10.7.

10.5 Let $\pi(i)$ be the shortest distance from the source of a network using arc costs p_{ij} as the lengths of the corresponding arcs. Would $\{ \pi(i) \}$ provide a useful set of vertex numbers for the application of algorithm (OOK)? Devise other rules for generating initial sets of vertex numbers.

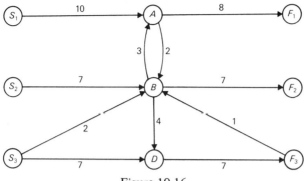

Figure 10.16

10.6 (de Braess Paradox) It is required to send 6 units of flow from S to F (see figure 10.17) in a most economical way, and in order to achieve this an equilibrium flow is sought. (A flow is in equilibrium if no reduction in cost is obtained by diverting an increment of flow from any path already used to any other path.) The cost is a non-linear function based on unit arc flow costs of

$$p_{Sa} = 2f_{Sa}, \quad p_{Sb} = f_{Sb} + 12, \quad p_{ab} = 2f_{ab} + \theta, \quad p_{aF} = f_{aF} + 12, \quad p_{bF} = 2f_{bF}$$

Verify that at an equilibrium the flow is

(1) evenly divided between the routes SaF and SbF with $C(\theta) = 126$ if $\theta > 9$, and
(2) evenly divided between the three routes SaF, $SabF$ and SbF with $C(\theta) = 132$ if $\theta = 2$.

Thus, reducing the cost of flow in ab leads to an increase in total cost $C(\theta)$! Indeed it follows that the insertion of an extra arc can increase the total cost of an equilibrium flow. This is known as the de Braess paradox and the above example is similar to one given in Fisk (1979) which in turn is based on the original one of de Braess.

10.7 Another paradox, this time for multi-commodity flows, is introduced in Fisk (1979). It is that increasing the flow in a network can lead to a reduction in total cost and is illustrated by the following example.

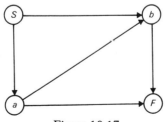

Figure 10.17

Consider a graph, on vertex set $\{a,b,c\}$ and with three arcs ab, ac and bc for which

a is the source for commodity 1, b is the sink and the flow is 3θ
a is the source for commodity 2, c is the sink and the flow is 20
b is the source for commodity 3, c is the sink and the flow is 51

Arc flow costs are $p_{ab} = f_{ab}$, $p_{bc} = f_{bc}$ and $p_{ac} = 40 + f_{ac}$ where f_{ij} is used to denote $\sum_{\alpha=1}^{3} f_{ij}^{\alpha}$. Suppose that the flow of commodity 2 is split at an equilibrium with $f_{ab}^{2} = f_{bc}^{2} = \xi$ and $f_{ac}^{2} = 20 - \xi$. Then the cost along both of the paths abc and ac must be the same (that is, $p_{ab} + p_{bc} = p_{ac}$). Verify that $\xi = 3 - \theta$ and the total cost is $C(\theta) = 3894 + 2\theta(3\theta - 11)$ if $0 \leqslant \theta < 3$. Thus $C(2) = 3874 < 3894 = C(0)$, and so increasing the flow of commodity 2 can decrease the total cost! What happens if the flow of commodity 1 ($= 3\theta$) exceeds 9?

10.8 Solve the TRP of example 10.12 starting from the given basis and with f_{61} as the initial incoming variable.

11 Heuristic Methods

When presented with a mathematical model problem P_M the problem-solver generally prefers to solve it exactly, that is to find a solution x_M^* guaranteed to be optimal. This is a natural attitude but there are several reasons why, in particular cases, it may be inappropriate to insist on an exact method.

(1) The model problem will almost always have been obtained by simplifying a real-life problem P_R in such a way that the essential features are retained. Consequently x_R^* the solution to P_R which 'corresponds' to x_M^*, may well not be optimal with respect to P_R. However it seems likely, if P_R has been well-modelled, that good solutions (to P_M) will correspond to good solutions (to P_R). It is then often preferable to spend the available effort on finding several good solutions to P_M, and hence to P_R, and presenting these to the decision maker who can then make the final choice bearing in mind the non-modelled (or neglected) features. On the other hand the decision-maker may use these solutions as starting points modifying them in the light of his experience.

(2) If an exact solution method is not readily available it may be that there is not time available to develop one.

(3) An exact method may require an inordinate amount of computer resources which is not justifiable compared to the extra benefit likely to be derived from having a guaranteed optimal solution.

(4) The amount of preparation and/or computer resources required may be such that an optimal solution cannot be obtained by the time it is required even though an exact method is available.

This is not to say that exact methods should not be developed and studied. A fund of exact methods to a wide spread of problems can alleviate difficulty (2) above, either by an exact method being on hand for solving P_M, or for solving a problem $P_{M'}$ which is sufficiently like P_M that P_R may reasonably be approximated by $P_{M'}$ (in this case an optimal solution to $P_{M'}$ will at least provide a benchmark against which other solutions may be compared). Also more efficient exact methods may alleviate difficulties (3) and (4) and perhaps lead to a better heuristic method [difficulty (1)].

Boffey (1976) lists four broad classes of heuristic method: improvement methods, constructive methods and decomposition methods which are dealt with in sections 11.1 to 11.3 respectively, and problem approximation mentioned above. Problem approximation is a familiar device, which in any case is involved in finding a model of a real-life situation, and will not be discussed further. It should be noted that this classification can of course be applied to exact methods as well (see exercises 11.1 and 11.3).

In this chapter more use is made of idealised problems (often TSP) as this makes easier the illustration of the heuristic concepts discussed. Finally it should be noted that the development of this chapter will be in terms of minimisation problems. (Maximisation problems can of course be treated by changing the sign of the objective.)

11.1 IMPROVEMENT METHODS

In Lin's method (section 7.2) a starting solution (tour) is selected and small adjustments made so as to lead to an improved solution (shorter tour). Another example of improving a solution in this way is provided by example 8.2. In both cases, each of the succession of adjustments must result in an improvement, and the adjustments are continued until no further improvement can be obtained by this means. These, and many others, belong to a class of heuristic methods called *improvement methods*. (They are also called hill-climb heuristics.) We now describe this class in formal terms and discuss strategies for implementing these methods.

We consider a problem P, with a finite set F of feasible solutions and with an objective function φ defined over F, for which a solution x^* is required with $\varphi(x^*) = \min_{x \in F} \varphi(x)$. Then, what is meant by a 'small' adjustment will depend on the problem solver who must specify a way of determining when a solution y is 'near' or 'a neighbour of' a solution x.

Definition 11.1
A function $N : F \rightarrow 2^F$, which associates a subset Nx with each $x \in F$, is a *neighbourhood function* over F if

(1) for all $x \in F$ $x \notin Nx$;
(2) $|Nx| \geq 1$.

Nx will be called the *neighbourhood* of x and $y \in F$ is a *neighbour* of $x \in F$ if $y \in Nx$. Clearly (F, N) can be thought of as a directed graph which we denote by $NG(P)$ and call a *neighbourhood graph* on F (see figure 11.1 for example). Most methods implicitly restrict themselves to a subgraph $SG(P)$ of $NG(P)$ which is defined below.

Definition 11.2
The *search graph*, $SG(P)$, for minimisation problem P is that graph (F, Γ) for which

$$\Gamma x = \{ y \mid y \in Nx \text{ and } \varphi(y) < \varphi(x) \}$$

Definition 11.3

A solution $\tilde{x} \in F$ is *locally optimal* (for a minimisation problem) if $\varphi(\tilde{x}) \leq \varphi(y)$ all $y \in Nx$; that is $\tilde{x}$ is a sink of the graph $SG(P)$. A solution $x^* \in F$ is *globally optimal* (for a minimisation problem) if $\varphi(x^*) \leq \varphi(x)$ for all $x \in F$.

Example 11.1 Four items A_0, A_1, A_2 and A_3 are to be grouped into two 'clusters' C_0 and C_1 so as to minimise the total separation between members of the same cluster, that is

minimise $\quad \varphi(x) = \varphi(x_0, x_1, x_2, x_3) = \frac{1}{2} \sum_{i=0}^{3} \sum_{j=0}^{3} d(A_i, A_j)[x_i x_j + (1-x_i)(1-x_j)]$

subject to $\quad 1 \leq \sum_{i=0}^{3} x_i \leq 3$

where $d(A_i, A_j)$ is some measure of the separation of A_i and A_j and $x_i = 0$ or 1 according as item A_i belongs to C_0 or C_1. Solve this problem for the case when the separation matrix is

$$\begin{bmatrix} 0 & 6 & 1 & 7 \\ 6 & 0 & 9 & 3 \\ 1 & 9 & 0 & 6 \\ 7 & 3 & 6 & 0 \end{bmatrix}$$

Solution Since the naming of the clusters is irrelevant we specify C_0 to be the cluster containing A_0, that is x_0 is fixed at 0.

The set of feasible solutions can thus be regarded as the set of 3-dimensional binary vectors

$$F = \{(x_1, x_2, x_3) \mid x_1 + x_2 + x_3 > 0\}$$

forming the vertices, other than the origin, of the 3-dimensional unit cube. We define a neighbourhood function by specifying that points (x_1, x_2, x_3) and (y_1, y_2, y_3) of F are neighbours if and only if they differ in precisely one co-ordinate, that is $|x_1 - y_1| + |x_2 - y_2| + |x_3 - y_3| = 1$. Then $NG(P)$ can be formed by joining pairs of points of F corresponding to an edge of the cube by a pair of oppositely directed arcs (figure 11.1a).

$SG(P)$ is now formed, for this particular choice of $NG(P)$, by deleting each arc xy for which $\varphi(x) \leq \varphi(y)$ (figure 11.1b).

The two sinks $e = (1, 0, 1)$ and $c = (0, 1, 1)$ correspond to local optima

$\quad\quad C_0 = \{A_0, A_2\}, \; C_1 = \{A_1, A_3\} \quad$ total separation 4
and $\; C_0 = \{A_0, A_1\}, \; C_1 = \{A_2, A_3\} \quad$ total separation 12

Clearly the first is both a local and global optimum whereas the second is local only.

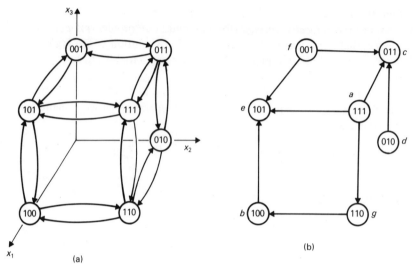

Figure 11.1 (a) and (b) show respectively the graphs $NG(P)$ and $SG(P)$ for the problem of example 11.1

It may also be noted that

$$C_0 = \{A_0\}, \quad C_1 = \{A_1, A_2, A_3\} \qquad \text{total separation } 18$$

is a global maximum and that

$$C_0 = \{A_0, A_1, A_3\}, \quad C_1 = \{A_2\} \qquad \text{total separation } 16$$
$$C_0 = \{A_0, A_1, A_2\}, \quad C_1 = \{A_3\} \qquad \text{total separation } 16$$

are local maxima.

Definition 11.4
Given a solution $s \in F$, a *hill-climb* (from *start point s*) is a sequence $s = x_0, x_1, x_2, \ldots, x_t = \tilde{x}$ of elements of F such that
(1) $x_i \in \Gamma x_{i-1}$ in $SG(P)$ $\quad i = 1, 2, \ldots, t$;
(2) $x_t = \tilde{x}$ is a local optimum.

Notice that since we are developing the theory for minimisation problems the hill-climbs will be 'climbs down' rather than 'climbs up'. A hill-climb can be viewed as starting at a vertex s of $SG(P)$ and forming a path $sx_1 x_2 \ldots \tilde{x}$ which ends at a sink and thus cannot be extended further. A *climbing strategy* **S** is, in general, required to specify the precise way in which these paths are formed. Given a partial path $x_0 x_1 \ldots x_{i-1}$, x_{i-1} not a sink, two obvious possibilities for strategies specifying the next point x_i are

(1) *random descent* $\mathbf{S}_{\text{rd}}$ with x_i chosen at random from Γx_{i-1}. [In practice one would use $\mathbf{S}'_{\text{rd}}$ with x_i being the first member of Γx_{i-1} (if any) encountered.]

Heuristic Methods

(2) *steepest descent* S_{sd} with x_i chosen at random from the subset of Γx_{i-1} comprising vertices minimising φ (over Γx_{i-1}).

Example 11.2 Analyse the chances of a single hill-climb producing a global optimum to the problem of example 11.1 if each $x \in F$ is equally likely to be chosen as start point. What are the chances of producing a global optimum from two hill-climbs?

Solution All possible hill-climbs, together with their probability of being chosen by S_{rd} and S_{sd}, are shown in table 11.1. Although true for this example it is not always the case that S_{sd} will give a higher probability of attaining a global optimum.

Table 11.1

	Probability of hill-climb (× 7)	
Hill-climb	S_{rd}	S_{sd}
ac	⅓	0
ae	⅓	1
agbe	⅓	0
be	1	1
c	1	1
dc	1	1
e	1	1
fc	½	0
fe	½	1
gbe	1	1
Probability of hill-climb to e	0.595 ($^{25}/_{42}$)	0.833 ($^{5}/_{7}$)

Steepest descent requires more computational effort (particularly near the beginning of a climb) and it is probably preferable, in general, to use random descent and use the effort saved to perform more hill-climbs.

In the case of two hill-climbs, S_{rd} and S_{sd} lead to the global optimum being attained (by at least one of the hill-climbs) with probabilities of 0.952 and 0.825 respectively if start points are constrained to be distinct. (If coincident start points are allowed the probabilities are reduced to 0.918 and 0.802). It is left as an exercise for readers to derive these probabilities.

Definition 11.5
The *catchment area* $S(\tilde{x})$ of any local optimum $\tilde{x} \in F$ is that set of elements of F from which there exists a hill-climb to $\tilde{x}$ using strategy S.

From table 11.1 it can be seen that for the problem of example 11.1

$S_{rd}(c) = \{a,c,d,f\}, \quad S_{rd}(e) = \{a,b,e,f,g\}$
$S_{sd}(c) = \{c,d\}, \quad S_{sd}(e) = \{a,b,e,f,g\}$

$S_{sd}(\tilde{x}) \subset S_{rd}(\tilde{x})$, and this is clearly true in general. It seems that it might be worth while imposing a condition that start points should not be too close to each other. This is found to be the case for the data of example 11.1 (see exercise 11.2), but will such tactics work for larger problems when small sets of start points are being used? This idea was pursued by Pursglove and Boffey (1980), but the experiments conducted were inconclusive as to whether it is worth while forsaking random selection of start points.

So far the discussion has centred on one particular form of neighbourhood function, and various other possibilities will now be introduced. First though we note that two broad classes of combinatorial problems are those for which the feasible solutions are naturally represented as binary n-vectors and those as permutations of n objects. Problems from the two classes will be termed *0–1 problems* and *permutation problems* respectively. Examples from the former class are 0–1 knapsack, shortest route, p-median and maximal flow problems; examples from the latter class are travelling salesman, linear and quadratic assignment, and flow-shop sequencing problems. (Of course any permutation problem can be reformulated as a 0–1 problem but we prefer nevertheless to think of the two classes as being distinct.)

A natural way of defining a neighbourhood function is to define a neighbourhood of $x \in F$ to be the set

$$N(x, \delta) = \{y \mid y \in F \ \& \ y \neq x \ \& \ d(x,y) \leq \delta\}, \quad \delta > 0$$

or some specified subset thereof, where d is a distance function defined over $NG(P)$. For a 0–1 problem the Euclidean distance $\sqrt{\{\Sigma(y_i - x_i)^2\}}$ could be taken, but since $(y_i - x_i)^2 = |y_i - x_i|$ the same neighbourhoods can be obtained using Hamming distances.

Definition 11.6
The *one-way distance* $d_o(x, y)$ from binary vector $x = (x_1, \ldots, x_n)$ to binary vector $y = (y_1, \ldots, y_n)$ is defined by

$$d_o(x,y) = \sum_{i=1}^{n} \max(y_i - x_i, 0)$$

The *Hamming distance* $d_H(x, y)$ between x and y is

$$d_H(x,y) = d_o(x,y) + d_o(y,x)$$

This leads to *Hamming neighbourhoods* $N_H(x;\delta)$ and (δ_1,δ_2)-neighbourhoods $N_R(x;\delta_1,\delta_2)$

$$N_H(x;\delta) = \{y \mid y \in F \ \& \ y \neq x \ \& \ d_H(x,y) \leq \delta\}$$
$$N_R(x;\delta_1,\delta_2) = \{y \mid y \in F \ \& \ y \neq x \ \& \ d_o(x,y) \leq \delta_1 \ \& \ d_o(y,x) \leq \delta_2\}$$

Heuristic Methods

For permutation problems we may define a distance function d_I such that $d_I(x, y) = \delta$ if y can be obtained from x by δ elementary operations of type I but not by $\delta - 1$ or fewer. Then a neighbourhood $N_I(x; \delta)$ of $x \in F$ is defined as

$$N_I(x; \delta) = \{ y \mid y \in F \ \& \ y \neq x \ \& \ d_I(x, y) \leq \delta \}$$

Some candidates for elementary operations are interchange of adjacent elements, interchanging positions of several elements, moving an element r places from position p ($p \in \{0, 1, \ldots, n-1\}$) to position $(p+r) \bmod(n)$ and at the same time shifting elements that were in positions $(p+1) \bmod(n) \ldots (p+r) \bmod(n)$ one place down.

Consider now two neighbourhood functions N_1 and N_2, for a particular problem, which satisfy $N_1(x) \subset N_2(x)$ all $x \in F$. Then $LO_1 \supset LO_2$ where LO_i is the set of local optima relative to N_i, $i = 1, 2$. Also if $\tilde{x}_1 \in LO_1 - LO_2$ there is a hill-climb from $\tilde{x}_1$ to a better solution $\tilde{x}_2 \in LO_2$ if N_2 is used. In this sense, larger neighbourhoods will lead to better quality solutions. However, this is offset by the extra effort required to search large neighbourhoods, and too large neighbourhoods should not be used. For example, using Lin's approach to TSPs or vehicle scheduling problems, neighbourhoods based on 2 or 3-transformations are usually employed. Also Roth (1970) recommends the use of $N_R(x; 2, 1)$ or $N_R(x; 3, 2)$.

Neighbourhoods can easily be too large for convenience; for example

$$| N_H(x; \delta) | = \binom{n}{1} + \binom{n}{2} + \ldots + \binom{n}{\delta}, \quad \delta \text{ a positive integer}$$

if all points of $N_H(x; \delta)$ also belong to F; this number is in excess of 100 000 if $n = 100$ and $\delta = 3$. Garfinkel and Nemhauser (1972) suggest using hill-climbs which use small neighbourhoods until a local optimum $\tilde{x}_1$ is obtained then larger neighbourhoods to improve x_1 to some new (and better) local optimum $\tilde{x}_2$. Another possibility is to search only a subset of $N(x)$ generated randomly.

Yet another approach is that of Lin and Kernighan (1973) (see also Kernighan and Lin, 1970) in which the neighbourhoods used may be regarded as data dependent in that points at distance 1 determine points searched at distance 2, etc.

Of course it is also important to choose neighbourhoods of an appropriate kind (exercise 11.3) and it is worth while expending some effort experimenting with different neighbourhood functions. Ones leading to long hill-climbs are likely to be better since they suggest larger catchment areas and hence fewer local optima.

11.2 CONSTRUCTIVE HEURISTIC METHODS

Improvement methods involve continually climbing from one feasible solution to another feasible solution; that is, the value of the objective function is continually improved while feasibility is maintained. Constructive methods, on the other hand, start from an infeasible 'solution' and move towards feasibility while keeping the value of the objective as good as possible. (The infeasibility may result from incomplete specification or violation of the constraints of the problem in some other way.)

Let C be a set of 'candidate' solutions for a problem P. Then a graph $CG(P)$ can be constructed with vertex set C and arcs xy where y is a neighbour of x. The neighbourhood functions allowed in this context are constrained by the requirement that for each arc xy, y should be 'nearer to feasibility' than x. Formalising this we define $CG(P)$ for a particular problem P as follows.

Definition 11.7
Let N be a neighbourhood function on candidate set C such that to each neighbour y of x there is a path from x to some element $z \in F$ which

(1) passes through y;
(2) has no more arcs than any other path from x to F.

The *construction graph* $CG(P)$ is then the graph on vertex set C which contains arc xy if and only if y is a neighbour of x.

Example 11.3 An example of a construction graph is given in figure 11.2 for a problem of selecting two items from four with y a neighbour of x if and only if they differ by only one component and the number of zero components of x is further from two than the number of zero components of y.

Definition 11.8
A *construction* from *start point* $s \in C$ is a path $s = x_0 x_1 x_2 \ldots x_t = \tilde{x}$ from s to $\tilde{x}$ where $\tilde{x} \in F$.

This says nothing about the way in which the construction should proceed. One way of producing feasible solutions from constructions is by using a real-valued function $h : C \rightarrow R$ to guide their formation.

Definition 11.9
For any function $h : C \rightarrow R$ an *h-construction* is a construction $x_0 x_1 \ldots x_t$ which also satisfies

$$h(x_{i+1}) = \min_{x_i y \in CG(P)} h(y) \quad \text{for } i = 0, \ldots, t-1$$

or

$$h(x_{i+1}) = \max_{x_i y \in CG(P)} h(y) \quad \text{for } i = 0, \ldots, t-1$$

Clearly, in the first case, it is likely that better solutions will be obtained if $h(y)$ tends to be smaller if there is a path from y to a 'good solution' in F. This prompts the definition of an approximation function.

Heuristic Methods 263

Definition 11.10
An *approximation* function [relative to a $CG(P)$] is a real-valued function $h : C \to R$ satisfying

(1) for all $x \in C$
[either $h(y) \geqslant h(x)$ or $h(y) \leqslant h(x)$ all y reachable from x];
(2) $h(x) = f(x)$ all $x \in F$.

Example 11.4 Find an approximation function for the problem of locating facilities at exactly p sites, from m available sites, so as to minimise total transport costs to n users. Illustrate its use by finding h-constructions for the specific data below if fixed costs at the various sites may be assumed to be all the same.

	Customer					
	1	2	3	4	5	
Site 1	18	5	5	19	10	
Site 2	12	7	8	10	8	$p = 2$
Site 3	10	15	11	8	9	$m = 4$
Site 4	14	6	4	10	22	$n = 5$
Customer wt	1	2	2	3	1	

Solution Since the fixed costs are the same for each site the problem becomes that of minimising total weighted transport costs f.
Choose the candidate set C to be the set of binary vectors with 4 coordinates. With the neighbourhood function N defined by neighbourhoods (section 11.1)

$$N(x) = N_R(x; 0, 1) \quad \text{if } x_1 + x_2 + x_3 + x_4 < 2$$
$$N(x) = N_R(x; 1, 0) \quad \text{if } x_1 + x_2 + x_3 + x_4 > 2$$

the construction graph $CG(P)$ is the one shown in figure 11.2 and as before the feasible set F is that subset of C for which

$$x_1 + x_2 + x_3 + x_4 = 2$$

A function $h : C \to R$ is now defined by

$h(x) = \infty$ if $x = 0$
$h(x) = $ total user cost if facilities located at sites i for which $x_i = 1$, when $x \neq 0$)

Clearly $h(x) = f(x)$ if $x \in F$. Also extra facilities cannot increase user costs and so $h(y) \leqslant h(x)$ if $y_i \geqslant x_i$, $i = 1, 2, 3, 4$. That is, h is an approximation function.

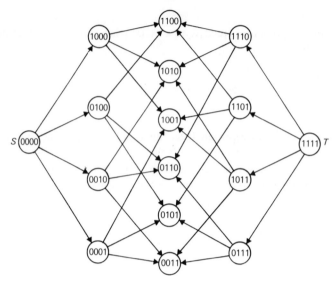

Figure 11.2

Suppose that we take $x = (0, 0, 0, 0)$ as start point for an h-construction $x_0 x_1 \ldots x_t$. Now

$N(x_0) = \{ (1,0,0,0), (0,1,0,0), (0,0,1,0), (0,0,0,1) \}$
$h[(1,0,0,0)] = 57$, $\quad h[(0,1,0,0)] = 45$, $\quad h[(0,0,1,0)] = 53$,
$h[(0,0,0,1)] = 56$

Since $(0, 1, 0, 0)$ yields the minimal value of $h(x)$ over $N(x_0)$, it is the next vertex x_1 of the h-construction.

$N(x_1) = \{ (1,1,0,0), (0,1,1,0), (0,1,0,1) \}$
$h[(1,1,0,0)] = 40$, $\quad h[(0,1,1,0)] = 41$, $\quad h[(0,1,0,1)] = 40$

As there is a tie for the $x \in N(x_1)$ minimising $h(x)$ we may select either $(1, 1, 0, 0)$ or $(0, 1, 0, 1)$ as the next vertex of the h-construction. Moreover, since $(1, 1, 0, 0)$ and $(0, 1, 0, 1)$ both belong to F the two possible h-constructions from $x_0 = (0, 0, 0, 0)$ are seen to be

$(0, 0, 0, 0) (0, 1, 0, 0) (1, 1, 0, 0)$
and $(0, 0, 0, 0) (0, 1, 0, 0) (0, 1, 0, 1)$

It is clearly seen that finding solutions by means of h-constructions is a heuristic method as $h[(1, 0, 1, 0)] = 37$ and $h[(0, 0, 1, 1)] = 37$.

Another possible choice of start point is $x_0 = (1, 1, 1, 1)$, and it is readily checked that

$N(x_0) = \{ (1,1,1,0), (1,1,0,1), (1,0,1,1), (0,1,1,1) \}$
$h[(1,1,1,0)] = 36$, $\quad h[(1,1,0,1)] = 39$, $\quad h[(1,0,1,1)] = 36$,
$h[(0, 1, 1, 1)] = 36$

There are four h-constructions

$(1,1,1,1)\,(1,1,1,0)\,(1,0,1,0)$
$(1,1,1,1)\,(1,0,1,1)\,(1,0,1,0)$
$(1,1,1,1)\,(0,1,1,1)\,(0,0,1,1)$
$(1,1,1,1)\,(1,0,1,1)\,(0,0,1,1)$

Thus it happened that for this problem a global optimum is found in each case.

As seen from the above example, constructions can lead to poor quality solutions being obtained. How might the quality be improved? We now outline some ways in which this might be approached. Clearly, for larger problems, a set $S = \{x^{(1)}, x^{(2)}, \ldots, x^{(r)}\}$ of start points can be used with the best feasible solution $\tilde{x}^{(i)}$ obtained being chosen as solution to the problem P. In example 11.3, r was 2 and

$\tilde{x}^{(1)} = (1,1,0,0)$ or $(0,1,0,1)$
$\tilde{x}^{(2)} = (1,0,1,0)$ or $(0,0,1,1)$

depending on how ties are broken. In this particular case $\tilde{x}^{(2)}$ is a better solution than $\tilde{x}^{(1)}$ (for all tie-breaking rules) with $g(\tilde{x}^{(2)}) = 37$.

Secondly, a solution x, obtained from a construction, can be refined by using an improvement method, the search now continuing over the set F with of course different neighbourhood functions (exercise 11.4). This combined constructive-improvement approach is a reasonable one to adopt if a single construction is being employed.

Thirdly, a certain amount of backtracking can be incorporated. If (for a minimisation problem) the guiding function is a lower bounding function (a special case of an approximation function) and backtracking performed as in B & B with cutoffs being made use of, then the method might be termed *heuristic B & B*. Heuristic B & B varies from a single construction to a B & B search tree being obtained according as no or 'complete' backtracking is used. An alternative way of backtracking in which an early decision can be changed without altering later ones, is given by Beale (1970).

The extent of backtracking may be controlled by allotting an amount of computational resources beforehand (for example Boffey and Hinxman, 1979). Alternatively, the procedure may be terminated when a certain minimal quality of solution is attained. This quality may be measured, in absolute or percentage terms, from the difference between the objective value for the incumbent solution (which provides an upper bound) and the value of the least lower bound for any active node on the search tree (which provides a lower bound). Graves and Whinston (1970) adopt this approach with the further refinement that bounds are determined probabilistically.

Lastly we might aim to improve quality of solution by improving the rank correlation between values $h(x)$ and objective values $f[y(x)]$ where y is a best feasible successor of x in $CG(P)$. This might be attempted by incorporating more problem specific information into the function h (compare the use of stronger bounds with B & B methods). Alternatively a function h_d may be *derived* from the guiding function h and itself used to guide constructions. h_d is defined recursively by

all $x \in C$ $h_1(x) = h(x)$
if $x \in F$ $h_d(x) = h(x) = f(x)$, $d > 0$

else $\begin{cases} h_d(x) = \min_{xy \in CG(P)} h(y) - (h \text{ decreasing towards feasibility}) \\ h_d(x) = \max_{xy \in CG(P)} h(y) - (h \text{ increasing towards feasibility}) \end{cases}$

Example 11.5   Solve the Euclidean distance TSP, on nine points, shown in figure 11.3a.

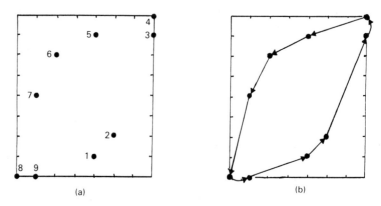

Figure 11.3 An optimal tour through the points 1 to 9 is shown in (b)

Solution   First note that since the problem is Euclidean the vertices on the convex hull will appear in order (theorem 8.3) in an optimal tour. Moreover, since the convex hull contains all the vertices, the tour 1234567891 (of length 23.719) must be an optimal tour.

Consider now that a part tour $\pi \equiv ijk \ldots m$ containing at most seven vertices has been built up, then $h(x)$ is defined to be the distance to the nearest vertex not yet on π; finally the tour is completed by returning to the start vertex. This method of obtaining tours by h-constructions is a nearest neighbour (NN) heuristic method.

Any vertex i may be selected as start vertex and corresponding constructions formed, as shown in table 11.2. In two cases, starting at vertices 1 and 9, alternative constructions exist and the particular one chosen would depend on the tie-breaking rule. It may be noted that the best solution obtained (start points 2, 6, 7, 9) has a length 3.3 per cent larger than the optimal tour, and the worst solution (obtained from start points 1 and 9) has a length 24.3 per cent larger that of the optimal tour. This shows how selecting the best solution obtained using several start points can lead to improvement.

Table 11.2

Heuristic	Start point	Tour	Value
Nearest neighbour	1	1298765341	27.097
	1	1276534981	29.482
	2	2198765342	24.496
	3	3456798123	24.557
	4	4356798124	25.334
	5	5679812345	24.557
	6	6798125346	27.258
	6	6534219876	24.496
	7	7653421987	24.496
	8	8912765348	29.151
	9	9876534219	24.496
	9	9812765349	29.482
Two-step	1	1987653421	24.496
	2	2198765342	24.496
	3	3456791283	32.496
	4	4356791284	33.064
	5	5346791285	33.645
	6	6532198746	32.917
	7	7653219847	36.275
	8	8912345678	23.719
	9	9123456789	23.719
	Optimal tour	1234567891	23.719

Since the problem is Euclidean a simple improvement method, to effect solution refinement, is to seek local optima, by removing crossings by means of 2-transformations (section 8.1). However, for this particular example all points lie on the convex hull and so the only 2-opt tour is 1234567891 (or its reverse), the global optimum.

We now look at the possibility of using *lookahead* to improve the quality of results. Specifically we shall consider looking ahead two steps by using a derived function h_2; this will be termed a two-step (TS) method. Since

$$\min_{j \neq i}(d_{ij} + \min_{k \neq i,j} d_{jk}) = \min_{\substack{j \neq i \\ k \neq i,j}} (d_{ij} + d_{jk})$$

a typical step consists of looking for two adjacent edges of minimal combined length which could be added (feasibly) to the part tour at vertex i. Only the first of these edges, ij, is actually added at this stage, then the process is repeated.

The tour is completed by adding the last two edges using the nearest neighbour approach.

We see from table 11.2 that TS leads to better results than NN for constructions from start points 1, 8 and 9 but to worse results for start points at vertices 3, 4, 5, 6 and 7 (the same solution is obtained for start point at vertex 2). While increasing the number of start points, refinement and backtracking cannot lead to a worsening in quality, the use of a derived function may. However, if we take all vertices as start points we have obtained an improvement for this particular problem and the global optimum has been obtained. It is interesting to note that the worst solution obtained using TS (52.9 per cent longer than the optimal tour) is actually much worse than the worst obtained using NN.

The above example has shown that a single construction can sometimes lead to solutions of very poor quality. This leads to the question of whether the percentage error can be bounded and to the search for methods for which this bound is near to zero (Fisher, 1980).

In order to illustrate the concepts of bounded error consider the 0–1 knapsack problem

$$\text{maximise} \quad \varphi = \sum_{i=1}^{n} p_i x_i \tag{11.1a}$$

subject to

$$\sum_{i=1}^{n} t_i x_i \leq T \tag{11.1b}$$

$$x_i \in \{0, 1\} \tag{11.1c}$$

For convenience we assume that the variables have been indexed so that

$$\frac{p_1}{t_1} \geq \frac{p_2}{t_2} \geq \ldots \geq \frac{p_n}{t_n}$$

and that $0 < t_i \leq T$, $0 < p_i$, for $1 \leq i \leq n$. Then a solution to the problem with constraint $0 \leq x_i \leq 1$ all i instead of $x_i \in \{0, 1\}$ is given by

$$x_1 = x_2 = \ldots = x_l = 1$$

$$x_{l+1} = (T - \sum_{i=1}^{l} t_i) / t_{l+1} < 1$$

$$x_{l+2} = \ldots = x_n = 0$$

If $l = n$ or $x_{l+1} = 0$ then this also provides a feasible solution to equation 11.1 which must hence be optimal. Even if $x_{l+1} > 0$ the solution

$$x_i = \begin{cases} 1 & \text{if } i \leq l \\ 0 & \text{if } i > l \end{cases}$$

Heuristic Methods

will often provide a good approximate solution. Denote the value of the solution obtained by this method $PMAX(0)$ and the value of an optimal solution by P^*.

Theorem 11.1
There exist 0–1 knapsack problems for all $n \geq 2$ for which $P^*/PMAX(0)$ is arbitrarily large.

Proof Let $\quad p_1 = 2, \quad t_1 = 1, \quad T = M > 2$

$$p_i = t_i = M, \quad 2 \leq i \leq n$$

where M is a positive integer. Clearly $P^* = M$ and $PMAX(0) = 2$. Since $P^*/PMAX(0) = M/2$ this ratio becomes arbitrarily large as M increases. □

This rather discouraging result illustrates very sharply how analysis of 'worst' cases can lead to very weak bounds even though the method yields quite good results in most cases encountered. That is, the worst-case bounds do not fully reflect the merits of a particular method. On the other hand if worst-case bounds are tight (near to 1) then we can assert that the method leads to good solutions in all cases.

Consider now a modification of the above heuristic method which essentially uses lookahead. Let S be a subset of the indices $\{1, 2, \ldots, n\}$ such that

$$x_i = \begin{cases} 1 & \text{if } i \in S \\ 0 & \text{if } i \notin S \end{cases}$$

provides a feasible solution to problem 11.1. This solution is now completed setting variables $x_i = 1, i \notin S$ in increasing order of i for as long as the constraint 11.1b is not violated. This leads to a value $\varphi(S)$ for φ. (Clearly $\varphi(S) = PMAX(0)$ if $S = \phi$.) Suppose further that we obtain such solutions for all subsets S for which $|S| = r$ and let

$$PMAX(r) = \max_{|S|=r} \varphi(S)$$

Theorem 11.2
For any 0–1 knapsack problem in n variables

$$\frac{P^*}{PMAX(r)} \leq 1 \left(\frac{1}{r+1}\right)$$

and the number of operations required to calculate $PMAX(r)$ is $O(n^{r+1})$.

Proof (Sahni 1975 and exercise 11.6).

This shows quite clearly how increased guaranteed accuracy may be achieved for the expenditure of extra effort. However, the 0–1 knapsack problem is

NP-hard and so it seems likely that in order to guarantee optimality an $O(a^n)$ algorithm is required (see appendix).

There are many studies on worst-case analysis: Ibarra and Kim (1975) and Lawler (1979) have given further results on the 0–1 knapsack problem; Garey and Johnson (1976) give a bibliography for results on various problems.

As noted above worst-case analysis may give poor results for algorithms which very often lead to good quality results. For example, the simplex algorithm has a very good record for solving large linear programming problems quickly and reliably even though the existence of families of LPs for which the running time grows exponentially can be contrived. Accordingly we now turn from trying to get a nearly optimal solution all the time to getting an optimal, or nearly optimal, solution, almost all the time. It is of course now necessary to introduce a probability distribution function f over the set of all possible problem instances. Though it is seldom known what probability distribution is realistic the approach does yield valuable insights not accessible through worst-case analysis (Karp, 1976).

Let C be a condition which, with probability q_N, is *not* satisfied for problem instances of size N whose frequency of occurrence is governed by probability density function f_N. Then the concept of 'almost all the time' is made more precise as follows.

Definition 11.11
Condition C is satisfied *almost everywhere (ae)* if $\sum_N q_N < \infty$.

Fisher and Hochbaum (1980) show that there is a polynomial time algorithm which solves the planar *p*-median problem, with constant weights and all vertices in the unit square, to within $1 + \epsilon$ almost everywhere. They divide the unit square into t^2 small squares $S_1, S_2, \ldots, S_{t^2}$ and let n_i be the number of vertices in S_i. If $d(S_i, S_j)$ is the distance between two nearest points with one in S_i and the other in S_j, then a related *m*-median problem, where $m = \min(p,$ number of occupied S_i), can be defined with each occupied square S_i now playing the role of vertex of weight n_i. Now

$$d(S_i, S_j) \leq d_{xy} \leq d(S_i, S_j) + \frac{2\sqrt{2}}{t} \tag{11.2}$$

if $x \in S_i$ and $y \in S_j$ (figure 11.4) and so since p squares are allocated to themselves

$$\varphi_{LB} \leq \varphi^* \leq \varphi_{LB} + \frac{2\sqrt{2}}{t}(n - p) \tag{11.3}$$

where φ_{LB} is a lower bound to φ^*, the value of a *p*-median, obtained by solving the related *m*-median problem.

A subset R containing p vertices is constructed by taking one vertex from each occupied S_i, $i = 1, 2, \ldots$ until $|R| = p$. (If $|R| < p$ after a vertex has been selected from every occupied S_i then the remaining vertices are selected arbitrarily.) The objective value $\varphi(R)$ for this choice satisfies (by equation 11.2)

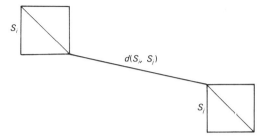

Figure 11.4

$$\varphi^* \leq \varphi(R) \leq \varphi_{LB} + \frac{2\sqrt{2}}{t}(n-p)$$

and hence by equation 11.3

$$|\varphi(R) - \varphi^*| \leq \frac{2\sqrt{2}}{t}(n-p) = \frac{2\sqrt{(2p)}}{\beta t}[\beta(\frac{n-p}{\sqrt{p}})]$$

can be made arbitrarily small for sufficiently large t. However, the larger t the more computation is involved. By careful analysis Fisher and Hochbaum (1980) show that

$$|\varphi(R) - \varphi^*| \leq \frac{2\sqrt{(2p)}}{\beta t}\varphi^* < \epsilon\varphi^*$$

provided $t \approx L\sqrt{p} = (2e^{\sqrt{(\pi p)}}/\epsilon)$ and prove that the computational effort required is no worse than

$$O[(\ln n)^{3/2} n^{1 + \ln L^2}]$$

Thus, given $\epsilon > 0$, the algorithm runs in a polynomially bounded time (see appendix), and finds a solution optimal to within $1 + \epsilon$ almost everywhere provided $p < \ln n$.

Algorithms for solving the Euclidean TSP and other problems to within $1 + \epsilon$ almost everywhere are given in Karp (1976).

11.3 PROBLEM REDUCTION : AND-OR GRAPHS

The reader may feel that many of the algorithms encountered so far are 'myopic' being based on local searching. The example below shows the application of a more global approach to a shortest route problem.

Example 11.6 Mr Brown wishes to travel by car from New York to Regina in Canada (figure 11.5). A major consideration is that the journey should be completed in a (relatively) short time. How might he choose his route?

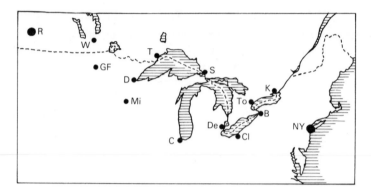

Figure 11.5 The Great Lakes Region of North America.
(B ≡ Buffalo, C ≡ Chicago, Cl ≡ Cleveland, D ≡ Duluth, De ≡ Detroit,
GF ≡ Grand Forks, K ≡ Kingston, M ≡ Minnesota, NY ≡ New York, R ≡ Regina,
S ≡ Sault Sainte Marie, T ≡ Thunder Bay, To ≡ Toronto, W ≡ Winnipeg)

Solution This is essentially a shortest route problem. A glance at a map reveals that the Great Lakes lie in the direct line from New York to Regina and three possibilities immediately suggest themselves

(1)   to travel south of the Lakes via Chicago; OR
(2)   to travel via Buffalo; OR
(3)   to travel north of the Lakes via Kingston.

Consider the second of these possibilities in which the problem of finding a route from New York to Regina is replaced by the two subproblems of finding routes

(4)   from New York to Buffalo; AND
(5)   from Buffalo to Regina.

Mr Brown, living in New York, knows the best route to Buffalo without consulting a map further; that is, NY–B (figure 11.6) is a *primitive* problem to which the answer is known or readily available. On the other hand the problem B–R is not so straightforward and two possibilities are

(6)   to travel south of Lake Michigan via Chicago; OR
(7)   north of Lake Huron via Sault Sainte Marie.

Node (7) has two successor problems the primitive B–S, and S–R. One possibility for the problem S–R is to travel south of Lake Superior via Duluth giving subproblems S–D and D–R. Note that D–R appears under the exploration of possibility (1) in which the journey is entirely south of the Lakes. Clearly problem D–R need be solved but once and this is the reason for joining node (8) to route D–R (figure 11.6). Other possibilities which might be considered by Mr Brown are also indicated by the AND—OR graph of figure 11.6. All the terminals of this graph represent primitive problems and so the graph [apart from solving subproblems such as (D—R) only once] is the *problem reduction* analogue of a complete enumeration tree.

Suppose now that on consideration of the primitive problems NY—B, B—S, NY—K and K—S it is found that it is quicker to travel to Sault Sainte Marie via Buffalo then, by the principle of optimality, any routes starting New York— Kingston—Sault Sainte Marie need not be further considered. Hence it follows that possibility (3) will not lead to an optimal route and so a cutoff is obtained. Thus it is seen that cycles are broken and a tree results, with a single route being represented by a subtree. For example, the subtree of figure 11.6 shown by the heavy lines represents the route New York—Boston—Cleveland—Chicago— Duluth—Grand Forks—Winnipeg—Regina.

Let $G = (X, \Gamma)$ be a directed graph without circuits and with a set of distinguished nodes $PRIM \subset X$ and another distinguished node $g \notin PRIM$. (The elements of $PRIM$ will be termed *primitive* nodes and g termed the *goal* node.) Suppose

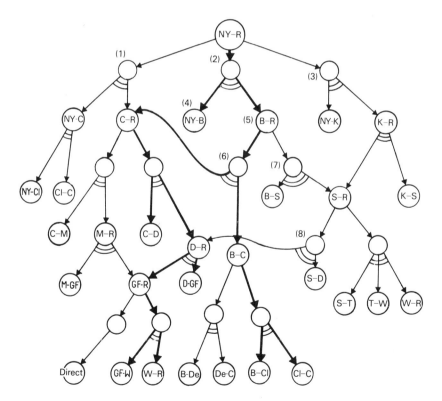

Figure 11.6 AND—OR graph for the routing problem of example 11.6. A solution subtree is distinguished by the thicker arcs

further that arcs are directed so that each arc is on a path from g to a node in PRIM. G together with a partition of $X - PRIM$ into two sets AND and OR will be termed an *AND–OR graph* (or an *AND–OR tree* if G is a tree). A *solution subtree* (V, T) is a tree with node set $V \subset X$ such that $g \in V$ and

(1) if $x \in V \cap $ AND then $Tx = \Gamma x$;
(2) if $x \in V \cap $ OR then $Tx \subset \Gamma x$ comprises a single element.

Figure 11.6 gave an example of an AND–OR graph and one of its solution subtrees. B & B and dynamic programming search trees can be regarded as special cases of AND–OR trees in which all nodes are OR-nodes (that is, AND = ϕ).

How should AND–OR graphs be searched? Consider the tree of figure 11.7.

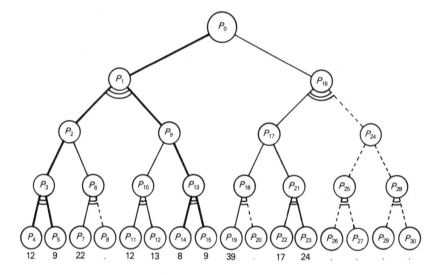

Figure 11.7 The optimal solution subtree is distinguished by the heavy lines

The best solution value $f(x)$ achievable from any OR node x is just the best solution value achievable from any immediate successor node (as is the case for B & B search trees in which all nodes are essentially OR nodes). That is

$$f(x) = \min_{y \in \Gamma x} f(y) \tag{11.4}$$

For an AND node x the equation is modified to

$$f(x) = \sum_{y \in \Gamma x} f(y) \tag{11.5}$$

Starting with terminals and working towards the root of the tree, applying either equation 11.4 or 11.5 as appropriate, the value of an optimal solution can be calculated.

Heuristic Methods

How might cutoffs be obtained? Suppose the graph is being searched in a depth-first manner and that the values of all primitive solutions are non-negative. Denote by $f_c(x)$ the current bound on $f(x)$ at node $x \notin PRIM$. $f_c(x)$ is determined as follows

At an OR node x, set $f_c(x) = \infty$ initially, and update by $f_c(x) \leftarrow \min [f_c(x), f(y)]$ when the value $f(y)$ becomes available [$f_c(x)$ is an upper bound to $f(x)$]

At an AND node x, set $f_c(x) = 0$ initially, and update by $f_c(x) \leftarrow f_c(x) + f(y)$ when the value $f(y)$ becomes available [$f_c(x)$ is a lower bound to $f(x)$]

Example 11.7 Find the cost of an optimal solution given the values $f(P_i)$ of primitive nodes shown against each terminal node P_i in figure 11.7.

Solution Initially $f_c(P_i) = \infty$ or 0 according as P_i is an OR-node or an AND-node.

The search proceeds via nodes P_0, P_1, P_2, P_3 to P_4. Then

$f(P_4) = 12$
$f_c(P_3) \leftarrow f_c(P_3) + f(P_4) = 0 + 12 = 12$
$f(P_5) = 9$
$f_c(P_3) \leftarrow f_c(P_3) + f(P_5) = 12 + 9 = 21$ [$=f(P_3)$ now]
$f_c(P_2) \leftarrow \min [f_c(P_2), f(P_3)] = \min (\infty, 21) = 21$
$f_c(P_6) = 0$
$f(P_7) = 22$
$f_c(P_6) \leftarrow f_c(P_6) + f(P_7) = 22$

Now since all costs are non-negative and P_6 is an AND-node $f(P_6) \geqslant f_c(P_6)$. Hence

$f(P_2) = \min [f_c(P_2), f(P_6)]$

$= \min (21, 22) = 21$

That is, $f(P_2)$ can be found without knowing the value of $f(P_8)$ and a cutoff is achieved. The search continues with

$f_c(P_1) \leftarrow f_c(P_1) + f(P_2) = 0 + 21 = 21$

then via $P_9, P_{10}, P_{11}, P_{10}, P_{12}, P_{10}, P_9, P_{13}, P_{14}, P_{13}, P_{15}$ and P_{13}, and $f(P_9)$ is found to be 17.

$f_c(P_1) \leftarrow f_c(P_1) + f(P_9) = 21 + 17 = 38$ [$=f(P_1)$ now]

$f_c(P_0) \leftarrow \min [f_c(P_0), f(P_1)] = \min (\infty, 38) = 38$

Proceeding via P_{16}, P_{17}, P_{18} to P_{19} we obtain $f(P_{19}) = 39$. Hence $f(P_{18}) \geqslant 39$.

Then
either $f(P_{21}) < f(P_{18})$ giving $f(P_{17}) = f(P_{21})$
or $f(P_{21}) \geq f(P_{18})$ giving $f(P_{16}) \geq f(P_{17}) = f(P_{18}) \geq 39$
In either case the value of $f(P_2)$ and $f(P_{18})$ is irrelevant since an upper bound of 38 has already been found for the OR-node P_0 — that is, a cutoff occurs. Continuing

$$f_c(P_{17}) \leftarrow 39$$
$$f(P_{21}) = 41$$
$$f(P_{17}) = \min\,[f_c(P_{17}), f(P_{21})\,] \geq \min\,(39, 41) = 39$$
$$f_c(P_{16}) \leftarrow f_c(P_{16}) + f(P_{17}) \geq 0 + 39 = 39$$

Now $38 = f(P_1) \geq f(P_0) = \min\,[f(P_1), f(P_{16})]$

$\qquad \geq \min\,[f(P_1), f_c(P_{16})]$
$\qquad = \min\,(38, 39) = 38$

Hence we can assert that $f(P_0) = 38$ and that an optimal solution is given by the subtree indicated by heavy lines in figure 11.7. The value of this solution is the sum of $f(P_4), f(P_5), f(P_{14})$ and $f(P_{15})$, that is, $12 + 9 + 8 + 9 = 38$.

Note that a lower bounding function may be available. It may well be, that for the expenditure of little effort, it can be asserted that $f(P_{16}) \geq 40$ in which case a 'stronger' cutoff could be achieved.

The precise formulation of the search procedure illustrated by means of the above example is left as an exercise (exercise 11.9). A similar situation which has been much studied (Nilsson, 1971, for example) arises in competitive situations with branches from OR-nodes corresponding to choices available to the decision maker and branches from AND-nodes corresponding to choices available to the 'opponent'. In this case $f_c(x)$ is initially set to $-\infty$ for AND-nodes and maximisation replaces summation.

The problem-reduction approach is probably used more in a heuristic context. For example, the list of choices available at an OR-node may be restricted heuristically or decomposition of a problem into subproblems may be approximate in some way.

Example 11.8 It is required to solve the Euclidean TSP on 25 points for which distances are given by figure 11.8.

Solution It is apparent that data points lie in four clusters (indicated by the letters A, B, C and D) and at a higher level fall into two groups $X = \{\,A, B\,\}$ and $Y = \{\,C, D\}$.

It is clear that any tour must cross between X and Y at least twice and that it is unlikely to make this relatively long crossing more than twice. Candidates for the crossings are 24–4 and either 12–3 or 12–25. This yields two alternative subproblems, the left-hand ones (figure 11.9) corresponding to tours which

Heuristic Methods

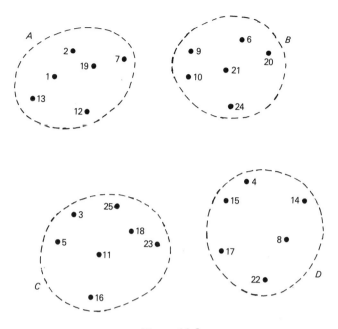

Figure 11.8

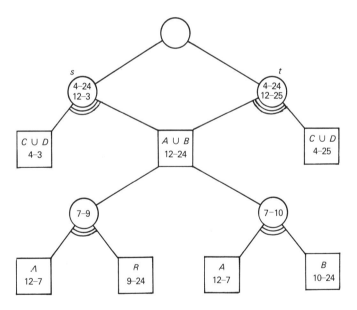

Figure 11.9 AND–OR graph for the problem of example 11.8. i–j in a circle indicates that i and j are joined directly. $i \underset{-}{\overset{P}{}} j$ in a box indicates that a shortest Hamilton path from i to j is sought in P

only cross the gap via 4—24 and 12—3. Other alternatives relating to crossings between $\{A, B\}$ and $\{C, D\}$ could have been included if desired.

Now in order to solve the problem at node s it is required to find a shortest Hamilton chain from vertex 12 to vertex 24 in X AND a shortest Hamilton chain from vertex 3 to vertex 4 in Y. Node s is thus being expressed as an AND-node. In a similar way t can be expressed as an AND-node with two subproblems one of which coincides with a subproblem of s.

Consider the Hamilton chain problem in $X = \{A, B\}$. The natural clustering suggests that the gap between A and B is crossed but once and probably via 7—9 or 7—10 creating two AND-nodes each of which splits naturally into two subproblems which might well be regarded as primitive and solved exactly.

The approach of the above example is particularly interesting in that it highlights how global information, readily available to a human, is being incorporated into the solution. A human is often better than a machine (computer program) at using global information and common sense, whereas the machine will usually be better at the level of detail. This suggests a combined man—machine approach in which the human approaches the problem 'from the top' and guides the machine to those (sub)problems which require the more routine use of detail. Krolak *et al.* (1971) did indeed put forward a man—machine method (more sophisticated than that sketched in the above example) for TSP. They claimed that the approach 'provides reasonable performance on very large problems (200 cities)'. Symmetric problems of this size are not being solved exactly by machine at the time of writing, though asymmetric TSPs of this size can be. It may not be long before symmetric problems involving 200 cities can be solved exactly entirely by machine and in a reasonably short time. This bears out the statement by Michie *et al.* (1968) that

> The man—machine combination should, we suggest, be regarded as a means of forcing an entry into territory which would be otherwise impenetrable, but which it is intended ultimately to subjugate to the arts of full mechanization. We thus see interactive techniques not as an end in themselves but as a step towards having the problem solved entirely by the machine behind the scenes.

Although the advantage may be passing to the machine as far as TSP is concerned (certainly for asymmetric problems) there are many other problems for which the state of the art is less advanced and where a man—machine approach is still appropriate. One reason why this may be so is that in order to make a model problem tractable certain features of the real problem may have been neglected. However, an experienced decision-maker could bear these in mind as he directs the problem-solving as part of a man—machine combination (for example, Boffey *et al.*, 1979).

Finally we note that Krolak and Nelson (1972) extend the approach of Krolak *et al.* (1971) to cover various urban problems including truck dispatching, school bussing and bussing for improving equal opportunity in schools.

EXERCISES

11.1 Discuss to which categories of heuristic or exact methods the following may be said to belong

(1) algorithm (CRASH);
(2) algorithm (CW);
(3) the SWEEP algorithm;
(4) the method of example 2.7;
(5) algorithm (K);
(6) the simplex method for solving LPs.

11.2 Using the data of example 11.1 four start sets $\{a\}$, $\{b, c\}$, $\{d, e\}$ and $\{f, g\}$ are generated by selecting pairs of opposite vertices of the unit cube with the origin being discarded if selected. For each set of start points find the probability that S_{rd} and S_{sd} will lead to the global optimum being attained. If each start set has equal chance of being chosen what is the overall probability that S_{rd} and S_{sd} will lead to the global optimum being attained?

11.3 It is required that n jobs be performed on a single machine. All jobs are ready at time zero when processing may begin. Each job i has an associated fixed processing time t_i and a time d_i at which the job is due to be completed. If processing on job i actually finishes at time f_i then the associated tardiness is $T_i = \max(0, f_i - d_i)$. The problem is to find a sequence of jobs [that is, a permutation of $(1, 2, \ldots, n)$] for which $\max_i T_i$ is minimised.

Suppose that an improvement method based on interchanging adjacent jobs (section 11.1) is being used. Prove that there is only one local optimum, which must thus be a global optimum.

11.4 The set of feasible solutions of a problem in 0–1 variables $x_1, \ldots, x_n$ is defined by the single constraint $x_1 + x_2 + \ldots + x_n = p$ where $0 < p < n$. A solution y is a neighbour of a solution x if and only if they differ in only two components ($x_k = 0, x_m = 1$ and $y_k = 1, y_m = 0$ for some $1 \leqslant k, m \leqslant n$).

Verify that if this neighbourhood function is used for the problem of example 11.4, then a global optimum is attained from every possible start point.

11.5 Use the neighbourhood function of exercise 11.4 to solve (heuristically) the problem of example 5.6.

11.6 Verify the result of theorem 11.2.

11.7 M identical machines $P_1, P_2, \ldots, P_M$ operate in parallel. n jobs $T_1, T_2, \ldots, T_n$ are scheduled, in turn, on to the machine which first becomes free giving a makespan of A (makespan is the finish time of the last job to finish minus the time of the start of processing of the first job). If

(1) a machine can process only one job at once;
(2) each job requires only one machine;
(3) each job T_i has a fixed processing time t_i;
(4) job T_k is the last job to finish being processed;

prove that

$$\sum_{i=1}^{n} t_i \geqslant t_k + M(A - t_k)$$

$$A \leqslant (2 - 1/M) A^*$$

where A^* is the minimum value of makespan under any schedule.

(LU 1977, abridged)

11.8 (Decision CPM—Crowston, 1970) The usual CPM network problem has been generalised by Crowston and Thompson to include the possibility of performing some jobs in several alternative ways. Each activity A_i is replaced by a set of activities $\sigma_i = \{ A_{i_1}, A_{i_2}, \ldots, A_{i_p} \}$ called a job set, and of these p alternative activities exactly one must be performed. Each activity A_α has an associated processing time t_α and an associated cost c_α. An example of decision CPM is given by the data of the table below.

Job set	Activities	Duration (days)	Cost (£)	Must be preceded by
1	1	5	110	—
2	2.1	7	85	1
	2.2	9	70	1
3	3	6	105	1
4	4	4	45	2.1, 2.2, 3
	5.1	5	130	3
5	5.2	8	65	3
	5.3	10	30	3
6	6.1	4	80	4, 5.1, 5.2, 5.3
	6.2	5	60	4, 5.1, 5.2, 5.3
7	3	6	115	2.1, 2.2, 5.3
8	8	10	70	2.1, 3
9	9	11	65	6.1, 6.2, 7, 8

There is a penalty of £30 per day by which PD exceeds 33 days.

(1) By searching an AND–OR tree maximising at AND nodes and minimising at OR nodes prove that the minimal project duration, irrespective of cost, is 33 days.

(2) By searching an AND–OR tree summing at AND nodes and minimising at OR nodes find the minimal cost, irrespective of time. (Note that after an activity j has been encountered for the first time its cost is set to zero to prevent double counting.)

Find the minimum overall cost (including lateness penalty) for $33 \leqslant T \leqslant 38$. This may be done as in (2) except that any path leading to $PD > T$ also leads to a cutoff (at the first OR node on this path at which the fact is discovered).

The interested reader may refer to Chapman and del Hoyo (1972) and Hindelang and Muth (1979) for further information.

11.9 Formalise a strategy for searching AND–OR trees which takes account of cutoffs as illustrated in example 11.7. Would such a strategy work for an AND–OR graph which is not a tree (Nilsson, 1971)?

Appendix: Computational Complexity

A simplified introduction to computational complexity will be given here. For a more extensive and rigorous treatment of this area, Garey and Johnson (1979) is particularly recommended.

In assessing the merits of an algorithm to solve a particular (type of) problem, $\wp$, various factors are of relevance. Some of these might be: how many instances (particular data sets) of $\wp$ will be involved; the sizes of these instances; 'characteristics' of the instances if known; the computing facilities available (speed, amount of main storage, etc.); the comprehensibility of the algorithm to the person applying it, and ease of implementation. Of course, if only a single small instance of $\wp$ is to be solved, then the most appropriate algorithm is likely to be the one with which the problem solver is most familiar. Of more concern is the case in which problem instances encountered are of a size near to the limits of computing feasibility (which usually means finding a solution in a not 'unreasonable' time). For this reason, performance of algorithms for large problem instances are of interest and the complexity of an algorithm is very often measured in terms of asymptotic behaviour. For this purpose it is convenient to use the 'order' notation. A positive function $f(n)$ is of *order* $g(n)$, written $f(n) = O(g(n))$, if $f(n)/g(n) \to c$ as $n \to \infty$ where c is a non-negative constant. Equivalently, $f(n) = O(g(n))$ if $f(n) < kg(n)$ all $n > N_0$ for some $k > 0, N_0 > 0$.

Example A.1
(1) $a = O(1)$ if $a > 0$ is a constant;
(2) $an^\alpha = O(n^\alpha)$ if $a > 0$ is a constant;
(3) $n^\alpha + n^\beta = O(n^\alpha)$ if $\alpha > \beta > 0$;
(4) $n^\alpha + a \log n = O(n^\alpha)$ if $\alpha > 0$ and a constant;
(5) $d + a \log n = O(\log n)$ if $a > 0$; a and d constant;
(6) $d^n + n^\alpha = O(d^n)$ if $d > 1$; all α;
(7) $n! + ad^n = O(n!)$.

These are standard results and for their justification the reader is referred to a textbook on advanced calculus. The asymptotic behaviour of the functions is *constant* for (1), *logarithmic* for (5), *polynomial* (power like) for (2), (3) and (4), *exponential* for (6) and *factorial* for (7).

Returning now to algorithms, the execution of an algorithm Å requires performance of certain elementary operations such as additions, multiplications and comparisons. If every elementary operation were to take the same time to perform (the *uniform-cost criterion*) then the execution time $t(Å)$ of $Å$ would be proportional to the number of elementary operations performed.

If an algorithm $Å$ can solve every problem instance of 'size' n in no more than $bn^\alpha + h(n)$ elementary operations, where $b > 0$ and $h(n)/n^\alpha \to 0$ as $n \to \infty$, then $Å$ is said to be *polynomially bounded*. ('*Exponentially bounded*', etc., may be defined in an analogous way.) Let P denote the set of problems for which polynomially bounded algorithms exist. Examples of problems belonging to P are

(1) Finding the maximum of a set of n numbers — this is straightforward using $n - 1$ comparisons and is clearly $O(n)$;
(2) Finding a minimal spanning tree (section 2.3) which can be achieved by algorithm (P) in $O(n^2)$ time;
(3) The assignment problem which can be solved by the algorithm of Dinic and Kronrod in $O(n^3)$ operations for an (n, n) cost matrix;
(4) Finding a maximal flow through a network. The method of Dinic–Karzanov solves this in $O(n^3)$ operations (section 9.2).

In the above examples the size of a problem was taken to be some natural parameter n. More strictly, the performance of an algorithm should be assessed by some problem-independent measure such as the number, m, of bits required to specify a problem: this would appear to involve the magnitude and accuracy of the data involved, but provided we are working to a fixed precision (single, double, ...) this is not a difficulty (Lawler, 1976). Also, if an algorithm is polynomial with respect to m then it is polynomial with respect to n but is not necessarily of the same order. (For example, with the MST problem, $½n(n-1)$ intervertex distances and the size n must all be specified giving $m = \beta[½n(n-1) + 1]$ where β is a constant depending on the precision involved. Thus algorithm (P) is $O(m)$ but may justifiably be described as $O(n^2)$.

Because of the rapid growth rate of the exponential function, problems which require exponential time for their solution are regarded as being 'intractable' while those which admit polynomial time solutions are 'tractable'. Of course, for small instances of a problem, an exponential time algorithm may be more effective than a polynomial algorithm. For example, (Aho *et al.*, 1974) $2^n < n^{10}$ for $2 < n < 59$, but 2^n rapidly gains thereafter $[2^{59} > 59^{10}$, from which it may be deduced that $2^n > (\frac{5}{3})^{n-59} n^{10}$ for $n > 59]$.

The asymptotic behaviour of an algorithm does however become more important as problem instances increase in size and as computers get faster (since larger problem instances can be solved)! For example, suppose a particular computer can solve, in one minute, any instance of a problem P up to size $n = 20$ using

(1) an $O(n)$ algorithm $Å_1$; and
(2) an $O(2^n)$ algorithm $Å_2$.

Then if the speed of the computer is increased by a factor of 8, problem instances of sizes up to $n = 160$ and $n = 23$ could be solved in 1 minute by $Å_1$ and $Å_2$ respectively. If the speed were to be increased by another factor of 8, then in 1 minute problem instances of sizes up to $n = 1280$ and $n = 26$ could be solved by $Å_1$ and $Å_2$ respectively. Further numerical examples illustrating the importance of the asymptotic behaviour of algorithms may be found in Aho *et al.* (1974).

Hitherto, algorithms considered have been *deterministic* and have been able to do only one thing at once. We shall now introduce the abstract concept of a *non-deterministic* algorithm in which operations may be performed in parallel. When a deterministic algorithm reaches a point at which there is a choice from a finite number, k, of alternatives then it proceeds by exploring a single alternative and must return later to explore the others. By contrast, a non-deterministic algorithm explores all k alternatives simultaneously. This is often described as the algorithm creating k copies of itself at the point at which the choice arises. This self-replication occurs instantaneously at every step where there is a choice among a finite set of alternatives. If any copy finds it has made an incorrect choice it stops executing. If any copy finds a solution then *all* copies stop executing.

Example A.2 The following is a non-deterministic algorithm to determine whether an n-city TSP with cost matrix (c_{ij}) has a tour of length at most b. It is similar to algorithm 9.2 in Reingold *et al.* (1977) but is expressed in a form more in keeping with the style we have adopted for deterministic algorithms. The statement $k \leftarrow$ choice (S) is to be interpreted as 'create $|S|$ copies of the algorithm at this point with each copy taking a different value for k from the set S'. Thus each copy of the algorithm will have its own values for j, k, *length* and S.

Algorithm

Step 1 (Setup)
    Set $k = 1$,   *length* = 0, $S = \{2, 3, \ldots, n\}$.
    Perform step 2 while S is non-empty.

Step 2 (Extend partial tour)
    Set $j \leftarrow k$, $k \leftarrow$ choice (S)
    length $\leftarrow$ *length* + c_{jk}, $S \leftarrow S - \{k\}$.

Step 3 (Termination)
    Set *length* $\leftarrow$ *length* + c_{k1}
    If *length* $\leq b$ then set *success = true*
    else set *success = false*.

The progress of this algorithm may be illustrated by the tree shown in figure A.1 which relates to $b = 18$ and the cost matrix

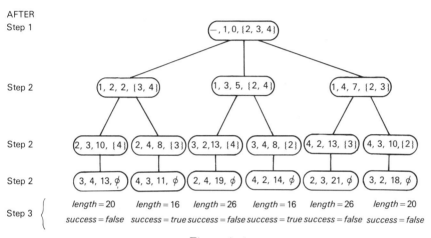

Figure A. 1

In this case the algorithm terminates successfully since (at least) one copy returns *success* = *true*.

Notice how backtracking is avoided by using the non-deterministic algorithm. Notice also that each application of step 2 will take the same time for each copy of the algorithm and so the 'running time' of this algorithm would be $O(n)$. *NP* will denote the set of problems for which a non-deterministic algorithm exists which will run in polynomial bounded time. Clearly a deterministic algorithm can be regarded as a special case of a non-deterministic algorithm and thus $P \subset NP$. It is not known whether $NP \neq P$ but the concensus of expert opinion is that this is likely to be so.

Suppose now that $\mathcal{P}$ and $\mathcal{P}_0$ are two problems and $Å_0$ is a non-deterministic algorithm for solving $\mathcal{P}_0$. Then $\mathcal{P}$ is said to *polynomially reduce* to $\mathcal{P}_0$ if there is a (non-deterministic) algorithm $Å$ to solve $\mathcal{P}$ such that

(1) to solve $\mathcal{P}$, $Å$ calls $Å_0$ at most $O(n^\alpha)$ times, where $\alpha > 0$;
(2) apart from calls on $Å_0$, algorithm $Å$ requires the performance of a number of elementary operations which is polynomial in the size of the instance of the $\mathcal{P}$ in question.

Now it is readily argued that if $\mathcal{P}$ polynomially reduces to $\mathcal{P}_0$ and $\mathcal{P}_0$ has a polynomial algorithm then so does $\mathcal{P}$. Thus $\mathcal{P}_0$ may be thought as being in a sense as hard as the problem $\mathcal{P}$. Indeed, any problem $\mathcal{P}'$ to which all problems in *NP* reduce is said to be *NP-hard*. An *NP*-hard problem which is also in *NP* is said to be *NP-complete*. We denote the set of *NP*-complete problems by *NPC*

$$NPC = \{ \mathcal{P}' | \mathcal{P}' \in NP \text{ and all } \mathcal{P} \in NP \text{ polynomially reduce to } \mathcal{P}' \} \subset NP$$

That *NPC* is non-empty was established by a celebrated result due to Stephen Cook which showed that it contained the *satisfiability* problem (Aho et al., 1974). Building on this result other problems have been shown to belong to *NPC* including the problems of determining the existence of

(1) a limited length tour (see example A.2);
(2) a Hamilton circuit in a graph; and
(3) a complete subgraph on k vertices in a graph;

and also the problem of partitioning a set of positive integers $\{ I_1, \ldots, I_n \}$ by finding a set $S \subset 1, \ldots, n$ such that $\sum_{i \in S} I_i = \sum_{i \notin S} I_i$, and many others, (Karp, 1975).

TSP has been shown to be *NP*-hard though it is not known whether it is in *NP* (and hence in *NPC*). Indeed the problem of finding an ϵ-optimal tour (section 11.2) for TSP is *NP*-hard (Reingold et al., 1977). Other *NP*-hard problems are finding a maximal complete subgraph of a graph (the 'clique problem'), the vehicle routing problem VRP (section 8.1), the p-median problem (section 5.2) and the quadratic assignment problem (chapter 1). For *NP*-hard problems, known deterministic algorithms are enumerative in nature and are effectively exponential. If any one *NP*-hard problem (complete or not) is polynomially bounded (with respect to a deterministic algorithm) then all *NP*-complete problems are. Thus $P \cap NPC = \phi$ or $P = NPC = NP$. Consequently, it is natural to design enumerative algorithms when dealing with problems in *NP*.

Whether a problem belongs to *P* or *NPC* has been used as a measure of *complexity*. It is a somewhat negative measure since it is based on 'worst-case' analysis. That is, inclusion of problem $\mathcal{P}$ in *NPC* is taken to imply that for any algorithm there are instances of $\mathcal{P}$ which cannot be solved in polynomial time. However, this does not preclude 'almost all' problem instances from being solved 'quickly'. For example, it has been shown that there is a class of linear programming problem instances for which the simplex algorithm requires an exponentially growing number of operations. However, in practice the amount of computation required grows polynomially with the number of variables and the number of constraints. From a practical point of view, the simplex algorithm is *effective*. There are difficulties with adopting this average-case approach when algorithms are being developed and tested, for it is often the case that new algorithms are tested on such a small number of (standard) problems that it is rather unsafe to generalise about their behaviour. On other occasions an algorithm is tested on a set of randomly generated data sets even though these may well be unrepresentative of the problems (data sets) likely to be encountered in practice; Karp (1976) introduced the idea of polynomially bounded algorithms which yield a solution of a guaranteed quality (such as being within 2 per cent of optimality) 'almost always'. This topic is discussed further in section 11.2.

For further reading on combinatorial optimisation and complexity, and for a guide to the relevant literature the survey article of Klee (1980) is recommended.

References

Aho, A.V., Hopcroft, J.E., and Ullman, J.D. (1974), *The Design and Analysis of Computer Algorithms* (Addison-Wesley, New York)

Arabeyre, J.P., Fearnley, J., Steiger, F.C., and Teather, W. (1969), The airline crew scheduling problem: A survey, *Transpn Sci.*, **3**, 140–63

Balas, E. (1965), An additive algorithm for solving linear programs with zero–one variables, *Ops Res.*, **13**, 517–46

Balas, E., and Ho, A. (1980), Set covering algorithms using cutting planes, heuristics, and subgradient optimization: a computational study, *Math. Programming Study*, **12**, 37–60

Barr, R.S., Glover, F., and Klingman, D. (1974), An improved version of the Out-of-Kilter method and a comparative study of computer codes, *Math. Programming*, **7**, 60–86

Barr, R.S., Glover, F., and Klingman, D. (1977), The alternating basis algorithm for assignment problems, *Math. Programming*, **13**, 1–13

Battersby, A. (1967), *Network Analysis for Planning and Scheduling* (Macmillan, London)

Beale, E.M.L. (1970), Selecting an optimum subset, in *Integer and Nonlinear Programming*, ed. J. Abadie (North-Holland, Amsterdam) pp. 451–62

Belford, P.C., and Ratliff, H.D. (1972), A network-flow model for racially balancing schools, *Ops Res.*, **20**, 619–28

Bellman, R. (1957), *Dynamic Programming* (Princeton University Press)

Benichou, M., Gauthier, J., Girodet, P., Hentges, G., Ribiere, G., and Vincent, D. (1971), Experiments in mixed integer linear programming, *Math. Programming*, **1**, 76–94

Berge, C. (1962), *Theory of Graphs and its Applications* (Methuen, London)

Bilde, O., and Krarup, J. (1977), Sharp lower bounds and efficient algorithms for the simple plant location problem, *Ann. Disc. Math.*, **1**, 79–97

Boffey, T.B. (1976), Heuristic methods for solving combinatorial problems, *London Papers in Regional Science*, **6**, 75–89

Boffey, T.B., and Hinxman, A.I. (1979), The optimal network problem, *Eur. J. O.R.*, **3**, 386–93

Boffey, T.B., and Karkazis, J. (1981), p-medians and multi-medians (In preparation)

Boffey, T.B., Edmond, E.D., Hinxman, A.I., and Pursglove, C.J. (1979), Routing container ships across the North Atlantic, *Opl Res. Q.*, **30**, 413–25

Bradley, G., Brown, G., and Graves, G.W. (1977), Designs and implementation

of large scale primal transshipment algorithms, *Mgmt Sci.*, **24**, 1–35
Breu, R., and Burdet, C-A. (1974), Branch and Bound experiments in zero–one programming, *Math. Programming Study*, **2**, 1–50
Burman, P.J. (1972), *Precedence Networks for Project Planning and Control* (McGraw-Hill, New York)
Burstall, R.M. (1967), Tree-searching methods with an application to a network design problem, in *Machine Intelligence*, vol. 1, ed. N.L. Collins and D. Michie (Oliver & Boyd, Edinburgh) pp. 65–85
Busacker, B.G., and Gowen, P.J. (1961), A procedure for determining a family of minimal-cost network flow patterns, *O.R.O. Technical Paper*, **15**
Business Week (1974), Making millions by stretching float, Nov. 23rd, 88–90
Buxey, S.M. (1979), The vehicle scheduling problem and route planning by Monte Carlo Simulation, *J. Opl Res. Soc.*, **30**, 563–73
Cassidy, P., and Bennett, B. (1972), TRAMP–A multi-depot vehicle dispatching problem, *Opl Res. Q.*, **23**, 151–63
Chang, Shi-Kuo (1972), The generation of minimal trees with a Steiner topology, *J. Ass. comput. Mach.*, **19**, 699–711
Chapman, C.B., and Del Hoyo, J. (1972), Progressive basic decision CPM, *Opl Res. Q.*, **23**, 345–59
Cheung, To-Yat (1980), Computational comparison of eight methods for the maximum network flow problem, *ACM Trans. mathl Software*, **6**, 1–16
Christofides, N. (1970), The shortest Hamiltonian chain of a graph, *SIAM J. appl. Math.*, **19**, 689–96
Christofides, N. (1972), Bounds for the travelling salesman problem, *Ops Res.*, **20**, 1044–56
Christofides, N. (1975), *Graph Theory: an Algorithmic Approach* (Academic Press, New York)
Christofides, N. (1976), The vehicle routing problem, *Rapp. a. Inst. Rech. Opl*, **10**, 55–70
Christofides, N., and Eilon, S. (1969), An algorithm for the vehicle dispatching problem, *Opl Res. Q.*, **20**, 309–29
Christofides, N., and Korman, S. (1975), A computational survey of methods for the set covering problem, *Mgmt Sci.*, **21**, 591–9
Christofides, N., and Viola, P. (1971), The optimum location of multi-centres of a graph, *Opl Res. Q.*, **22**, 145–54
Church, R., and ReVelle, C.S. (1974), The maximal covering location problem, *Pap. Reg. Sci. Ass.*, **32**, 101–12
Clarke, G., and Wright, J. (1963), Scheduling of vehicles from a central depot to a number of delivery points, *Ops. Res.*, **11**, 568–81
Corkindale, D.R. (1975), Queueing theory in the solution of a transport evaluation problem, *Opl Res. Q.*, **26**, 259–71
Cornuejols, G., Fisher, M.L., and Nemhauser, G.L. (1977), Location of bank accounts to optimize float, *Mgmt Sci.*, **21**, 789–810
Courant, R., and Robbins, H. (1941), *What is Mathematics?* (Oxford University

Press, New York)

Crowston, W. (1970), Decision CPM: network reduction and solution, *Opl Res. Q.*, **21**, 435–52

Cunto, E. (1978), Scheduling boats to sample oil wells in Lake Maracaibo, *Ops Res.*, **26**, 183–96

Daganzo, C.F. (1977a), On the traffic assignment problem with flow dependent costs – I, *Transp. Res.*, **11**, 433–7

Daganzo, C.F. (1977b), On the traffic assignment problem with flow dependent costs – II, *Transp. Res.*, **11**, 439–41

Dakin, R.J. (1965), A tree search algorithm for mixed integer programming problems, *Comput. J.*, **8**, 250–5

Dantzig, G.B. (1963), *Linear Programming and Extensions* (Princeton University Press)

Dantzig, G.B. (1967), All shortest routes in a graph, *Proceedings of the International Symposium on Theory of Graphs (Rome, 1966)* (Gordon & Breach, New York) pp. 91–2

Dantzig, G.B., Blattner, W.O., and Rao, M.R. (1967), Finding a cycle in a graph with minimal cost to time ratio with applications to a ship routing problem. *Proceedings of the International Symposium on Theory of Graphs (Rome, 1966)* (Gordon & Breach, New York) pp. 77–83

Denardo, E.V., and Fox, B.L. (1979), Shortest route methods: 1. Reaching, pruning and buckets, *Ops Res.*, **27**, 161–86

Deo, N. (1976), Note on Hopcroft and Tarjan planarity algorithm, *J. Ass. comput. Mach.*, **23**, 74–5

Desler, J.F., and Hakimi, S.L. (1969), A graph-theoretic approach to a class of integer programming problems, *Ops Res.*, **17**, 1017–33

Dial, D., Glover, F., Karney, D., and Klingman, D. (1977), A computational analysis of alternative algorithms and labelling techniques for finding shortest path trees, *Research Report CCS 291* (Center for Cybernetic Studies, University of Texas, Austin, Texas)

Dijkstra, E.W. (1959), A note on two problems in connection with graphs, *Num. Math.*, **1**, 269–71

Dinic, E.A. (1970), Algorithm for solution of a problem of maximum flow in a network with power estimation, *Soviet Math. Dokl.*, **11**, 1277–80

Dinic, E.A., and Kronrod, M.A. (1969). An algorithm for the solution of the assignment problem. *Soviet Math. Dokl.*, **10**, 1324–6 (English translation)

Dionne, R., and Florian, M. (1979), Exact and approximate algorithms for optimal network design, *Networks*, **9**, 37–59

Dreyfus, S.E. (1969), An appraisal of some shortest path algorithms, *Ops Res.*, **17**, 395–412

Drinkwater, R.W. (1977), Placing of multiplexers to minimise cable distance, *Opl Res. Q.*, **28**, 267–73

Eastman, W.L. (1958), Linear Programming with Pattern Constraints, Ph.D. thesis, Harvard University

Edmonds, J., and Karp, R.M. (1972), Theoretical improvements in algorithmic efficiency for network flow problems, *J. Ass. comput. Mach.*, **19**, 248–64

Eilon, S., Watson-Gandy, C.D.T., and Christofides, N. (1971), *Distribution Management: Mathematical Modelling and Practical Analysis* (Griffin, London)

Elmaghraby, S.E. (1977), *Activity Networks: Project Planning and Control by Network Models* (Wiley, New York)

Elshafei, A.M. (1977), Hospital layout as a quadratic assignment problem, *Opl Res. Q.*, **28**, 167–79

Elzinga, J., and Hearn, D.W. (1972), Geometrical solutions for some minimax location problems, *Transp. Sci.*, **6**, 379–94

Erlenkotter, D. (1979), A dual-based procedure for uncapacitated facility location, *Ops Res.*, **26**, 992–1009

Fisher, M.L. (1980), Worst-case analysis of heuristic algorithms, *Mgmt Sci.*, **26**, 1–17

Fisher, M.L., and Hochbaum, D.S. (1980), Probabilistic analysis of the planar K-median problem. *Math. Ops Res.*, **5**, 27–34

Fisk, C. (1979), More paradoxes in the equilibrium assignment problem, *Transp. Res.*, **13B**, 305–9

Floyd, F.W. (1962), Algorithm 97: shortest path, *Communs Ass. comput. Mach.*, **5**, 345

Folie, M., and Tiffin, J. (1976), Solution of a multi-product manufacturing and distribution problem, *Mgmt Sci.*, **23**, 286–96

Ford, L.R. jr (1956), Network Flow Theory, Rand Corporation Report P-923.

Ford, L.R. jr, and Fulkerson, D.R. (1962), *Flows in Networks* (Princeton University Press)

Foster, B.A., and Ryan, D.M. (1976), An integer programming solution to the vehicle scheduling problem. *Opl Res. Q.*, **27**, 367–84

Francis, R.L., and White, J.A. (1974), *Facility Layout and Location: an Analytical Approach* (Prentice-Hall, Englewood Cliffs, N.J.)

Frank, H., and Frisch, I.T. (1971), *Communication, Transportation and Flow Networks* (Addison-Wesley, Reading, Mass.)

Garey, M.R., and Johnson, D.S. (1976), Approximation algorithms for combinatorial problems: an annotated bibliography, in *Algorithms and Complexity: New Directions and Recent Results*, ed. J.F. Traub (Academic Press, New York) pp. 41–52

Garey, M.R., and Johnson, D.S. (1979), *Computers and Intractability – A guide to the theory of NP-completeness* (Freeman, San Francisco)

Garfinkel, R.S., and Nemhauser, G.L. (1972), *Integer Programming* (Wiley, New York)

Garfinkel, R.S., Neebe, A.W., and Rao, M.R. (1977), The m-center problem: minimax facility location, *Mgmt Sci.*, **23**, 1133–42

Gaskell, T. (1967), Bases for vehicle fleet scheduling, *Opl Res. Q.*, **18**, 281–95

Geoffrion, A.L. (1974), Lagrangean relaxation, *Math. Progm. Stud.*, **2**, 82–114

Gilett, B., and Miller, L. (1974), A heuristic algorithm for the vehicle dispatch problem, *Ops Res.,* **22**, 340—9

Glover, F., and Klingman, D. (1973), A note on computational simplifications in solving generalized transportation problems, *Transp. Sci.,* **7**, 351—61

Glover, F., Karney, D., and Klingman, D. (1972), The Augmented Predecessor Index Method for locating stepping stone paths and assigning dual prices in distribution problems, *Transp. Sci.,* **6**, 171—80

Glover, F., Klingman, D., and Stutz, J. (1973), Extensions of the Augmented Predecessor Index Method to generalized network problems, *Transp. Sci.,* **7**, 377—84

Glover, F., Klingman, D., and Stutz, J. (1974), The Augmented Threaded Index Method for network optimization, *INFOR,* **12**, 293—8

Glover, F., Hultz, J., Klingman, D., and Stutz, J. (1978), Generalized networks: a fundamental computer-based planning tool, *Mgmt Sci.,* **24**, 1209—20

Golden, B.L. (1976), Shortest-path algorithms: a comparison, *Ops Res.,* **24**, 1164—8

Golden, B.L., Magnanti, T.L., and Nguyen, H.Q. (1977), Implementing vehicle routing algorithms, *Networks,* **7**, 113—48

Goldman, A.J. (1969), Optimum locations for centers in a network, *Transp. Sci.,* **4**, 352—60

Goldman, A.J. (1972), Minimax location of a facility on a network, *Transp. Sci.,* **6**, 407—18

Gower, J.C., and Ross, G.J.S. (1969), Minimum spanning trees and single linkage cluster analysis, *Appl. Statists,* **18**, 54—64

Graves, G.W., and Whinston, A.B. (1970), An algorithm for the Quadratic Assignment Problem, in *Integer and Nonlinear Programming*, ed. J. Abadie (North Holland, Amsterdam) pp. 473—97

Hakimi, S.L. (1964), Optimum locations of switching centers and the absolute centers and medians of a graph, *Ops Res.,* **12**, 450—9

Hakimi, S.L. (1965), Optimal distribution of switching centres in a communications network and some related graph theoretic problems, *Ops Res.,* **13**, 462—75

Hakimi, S.L. (1971), Steiner's problem in graphs and its implications, *Networks,* **1**, 113—33

Hakimi, S.L., Schneidel, E.F., and Pierce, J.G. (1978), On p-centres in networks, *Transp. Sci.,* **12**, 1—15

Halpern, J. (1976), The location of a center-median convex combination on an undirected tree, *J. Reg. Sci.,* **16**, 237—45

Handler, G.Y. (1973), Minimax location of a facility in an undirected tree graph, *Transp. Sci,* **7**, 287—93

Handler, G.Y. (1978), Finding two-centres of a tree: the continuous case, *Transp. Sci.,* **12**, 93—106

Handler, G.Y., and Mirchandani, P.B. (1979), *Location on Networks, Theory and Algorithms* (MIT Press, Cambridge, Mass.)

Hart, P., Nilsson, N., and Raphael, B. (1968), A formal basis for the heuristic determination of minimum cost paths, *IEEE Trans. Syst. Man Cybernet.*, **4**, 100–7

Hansen, K.H., and Krarup, J. (1974), Improvements of the Held and Karp algorithm for the symmetric travelling salesman problem, *Math. Programming*, **4**, 87–98

Held, M., and Karp, R.M. (1970), The travelling salesman problem and minimum spanning trees, *Ops Res.*, **18**, 1138–62

Held, M., and Karp, R.M. (1971), The travelling salesman problem and minimum spanning trees, Part II, *Math. Programming.*, **1**, 6–25

Hindelang, T.J., and Muth, J.F. (1979), A dynamic programming algorithm for Decision CPM networks, *Ops Res.*, **27**, 225–41

Holmes, R.A., and Parker, R.G. (1976), A vehicle scheduling procedure based upon savings and a solution perturbation scheme, *Opl Res. Q.*, **27**, 83–92

Hopcroft, J., and Tarjan, R. (1974), Efficient planarity testing, *J. Ass. comput. Mach.*, **21**, 549–68

Horne, G.J. (1980), Finding the K least cost paths in an acyclic activity network, *J. Opl Res. Soc.*, **31**, 443–8

Hu, T.C. (1969), *Integer Programming and Network Flows* (Addison-Wesley, Reading, Mass.)

Ibarra, O.H., and Kim, C.E. (1975), Fast approximation algorithms for the Knapsack and sum of subset problems, *J. Ass. comput. Mach.*, **22**, 463–8

Ignall, E., and Schrage, L.E. (1965), Applications of the Branch and Bound technique to some flow shop scheduling problems, *Ops Res.*, **13**, 400–12

Johnson, D.B. (1973), A note on Dijkstra's shortest path algorithm, *J. Ass. comput. Mach.*, **20**, 385–8

Johnson, E.L. (1966), Networks and basic solutions, *Ops Res.*, **14**, 619–23

Karp, R.M. (1975), On the computational complexity of combinatorial problems, *Networks*, **5**, 45–68

Karp, R.M. (1976), The probablistic analysis of some combinatorial search algorithms, in *Algorithms and Complexity: New Directions and Recent Results*, ed. J.F. Traub (Academic Press, New York) pp. 1–19

Karzanov, A.V. (1974), Determining the maximal flow in a network by the method of preflows, *Soviet Math. Dokl.*, **15**, 434–7

Kaufmann, A. (1967), *Graphs, Dynamic Programming and Finite Games* (Academic Press, New York)

Kernighan, B.W., and Lin, S. (1970), An efficient heuristic procedure for partitioning graphs, *Bell Syst. tech. J.*, **49**, 291–307

Klee, V. (1980), Combinatorial optimization: what is the state of the art?, *Math. Ops Res.*, **5**, 1–26

Klein, M. (1963), On assembly line balancing, *Ops Res.*, **11**, 274–81

Klein, M. (1967), A primal method for minimal cost flows, with application to the assignment and transportation problems, *Mgmt Sci.*, **14**, 205–20

Klingman, D., Napier, A., and Stutz, J. (1974), NETGEN – A program for generating large scale (un)capacitated assignment, transportation and minimum cost flow network problems, *Mgmt Sci.*, **20**, 814–21

Knuth, D.E. (1973), *The Art of Computer Programming Vol. 3, Sorting and Searching* (Addison-Wesley, Reading, Mass.)

Kolesar, P.J. (1966), A Branch and Bound Algorithm for the Knapsack Problem, *Mgmt Sci.*, **13**, 723–35

Krarup, J., and Pruzan, P.M. (1977), Selected families of discrete location problems. Part I: Centre problems. Part II: Median problems. Rapport nr. 77/7., University of Copenhagen.

Krolak, P.D., and Nelson, J.H. (1972), A man–machine approach for creative solutions to urban problems, in *Machine Intelligence*, **7**, ed. B. Meltzer and D. Michie (Edinburgh University Press) pp. 241–66

Krolak, P., Felts, W., and Marble, G. (1971), A man–machine approach toward solving the travelling salesman problem, *Communs Ass. comput. Mach.*, **14**, 327–34

Kruskal, J.B. (1956), On the shortest spanning subtree of a graph and the travelling salesman problem, *Proc. Am. math. Soc.*, **7**, 48–50

Kuratowski, K. (1930), Sur le problème des courbes gauches en topologie, *Fundam. Math.*, **15**, 271–83

Kwan, M-K. (1962), Graphic programming using odd or even points, *Chin. Math.*, **1**, 273–7

Land, A.H., and Doig, A.G. (1960), An automatic method of solving discrete programming problems, *Econometrica*, **28**, 497–520

Lawler, E.L. (1976), *Combinatorial Optimization, Networks and Matroids* (Holt–Rinehart–Winston, New York)

Lawler, E.L. (1979), Fast approximation algorithms for Knapsack Problems, *Math. Ops Res.*, **4**, 339–56

Lawler, E.L., and Bell, M.D. (1967), A method for solving discrete optimization problems, *Ops Res.*, **15**, 1098–112

LeBlanc, L.J., Morlok, E.K., and Pierskalla, W.P. (1975), An efficient approach to solving the road network equilibrium traffic assignment problem, *Transp. Res.*, **9**, 309–18

Lin, S. (1965), A computer solution of the travelling salesman problem, *Bell Syst. tech. J.*, **44**, 2245–69

Lin, S., and Kernighan, B.W. (1973), An effective heuristic for the Travelling Salesman Problem, *Ops Res.*, 498–505

Little, J.D.C., Murty, K.G., Sweeney, D.W., and Karel, C. (1963), An algorithm for the travelling salesman problem, *Ops Res.*, **11**, 979–89

Lock, D. (1971), *Industrial scheduling* (Gower, London)

Malhotra, V.M., Kumer, M.P., and Maheshwari, S.N. (1978), An $O(|V|^3)$ algorithm for finding maximum flows in networks, *Inf. Process. Lett.*, **7**, 277–8

Mandl, C. (1979), *Applied Network Optimization* (Academic Press, New York)

Marshall, C.W. (1971), *Applied Graph Theory* (Wiley Interscience, New York)

Michie, D., Fleming, T.G., and Oldfield, J.V. (1968), A comparison of heuristic, interactive, and unaided methods of solving a shortest route problem, in *Machine Intelligence*, vol. 3, ed. D. Michie (Edinburgh University Press) pp. 245–55

Minieka, E. (1978), *Optimization Algorithms for Networks and Graphs* (Marcel Dekker, Basel)

Moder, J.J., and Phillips, C.R. (1970), *Project Management with CPM and PERT*, 2nd ed. (Rheinold, New York)

Mole, R.H. (1979), A survey of local delivery vehicle routing methodology, *J. Opl Res. Soc.*, **30**, 245–52

Mole, R.H., and Jameson, S.R. (1976), A sequential route-building algorithm employing a generalised saving criterion, *Opl Res. Q.*, **27**, 503–11

Moore, P.G., and Thomas, H. (1973), The rev counter decision, *Opl Res. Q.*, **24**, 337–51

Nauss, R.M. (1978), An improved algorithm for the capacitated facility location problem, *J. Opl Res. Soc.*, **29**, 1195–1201

Nemhauser, G.L. (1972), A generalized permanent label setting algorithm for the shortest path between all nodes, *J. math. Analysis Applic.*, **38**, 328–34

Nilsson, N.J. (1971), *Problem-solving Methods in Artificial Intelligence* (McGraw-Hill, New York)

Padberg, M.W., and Hong, S. (1980), On the symmetrical travelling salesman problem: a computational study, *Math. Programming Stud.*, **12**, 78–107

Pohl, I. (1977), Improvements to the Dinic–Karzanov network flow algorithm, Technical Report No. HP–77–11–001, (Dept of Information Sciences, University of California, Santa Cruz)

Prim, R.C. (1957), Shortest connection networks, and some generalizations, *Bell Syst. tech. J.*, **36**, 1389–401

Pursglove, C.J., and Boffey, T.B. (1980), Improvement methods: how should starting solutions be chosen? *Math. Programming Stud.*, **13**, 135–42

Rand, G.K. (1976), Methodological choices in depot locational studies, *Opl Res. Q.*, **27**, 241–9

Reddi, S.S., and Ramamoorthy, C.V. (1973), A scheduling problem, *Opl Res. Q.*, **24**, 441–6

Reingold, E.M., Nievergelt, J., and Deo, N. (1977), *Combinatorial Algorithms: Theory and Practice* (Prentice-Hall, Englewood Cliffs, N.J.)

ReVelle, C.S., and Church, R. (1977), A spatial model for the location construct of Teitz, *Pap. Reg. Sci. Ass.*, **39**, 129–35

ReVelle, C.S., Marks, D., and Liebman, J.C. (1970), An analysis of private and public sector location models, *Mgmt Sci.*, **16**, 692–707

Roth, R.H. (1970), An approach to solving linear discrete optimization problems, *J. Ass. comput. Mach.*, **17**, 303–13

Russell, R.A. (1977), An effective heuristic for the M-tour travelling salesman problem with some side conditions, *Ops. Res.*, **25**, 517–24

Saha, J.L. (1970), An algorithm for bus scheduling, *Opl Res. Q.*, **21**, 463–74

Sahni, S. (1975), Approximate algorithms for the 0–1 knapsack problem,

J. Ass. comput. Mach., **22**, 115–24

Salkin, H.M. (1975), *Integer Programming* (Addison-Wesley, Reading, Mass.)

Schrage, L.E. (1970), Solving resource-constrained network problems by implicit enumeration – nonpreemptive case. *Ops Res.*, **18**, 263–78

Scott, A.J. (1971), *Combinatorial Programming, Spatial Analysis and Planning* (Methuen, London)

Segal, M. (1974), The operator scheduling problem: a network flow approach, *Ops Res.*, **22**, 808–23

Seppänen, J., and Moore, J.M. (1970), Facilities planning with graph theory, *Mgmt Sci.*, **17**, B242–53

Shih, W. (1979), A Branch and Bound method for the multi-constraint zero–one Knapsack Problem, *J. Opl Res. Soc.*, **30**, 369–78

Spira, P.M. (1973), A new algorithm for finding all shortest paths in a graph of positive arcs in average time $O(n^2 \log \log n)$, *SIAM J. Comput.*, **2**, 28–32

Srinivasan, V., and Thompson, G.L. (1972), Accelerated algorithms for labelling and relabelling trees with applications for distribution problems, *J. Ass. comput. Mach.*, **19**, 712–26

Taha, H.A. (1975), *Integer Programming: Theory, Algorithms and Computations* (Academic Press, New York)

Taha, H.A. (1976), *Operations Research, an Introduction*, second edition (Macmillan, New York)

Taylor, A.J. (1976), Systems dynamics in shipping, *Opl Res. Q.*, **27**, 41–56

Tillman, F.A., and Cain, T. (1972), An upper bounding algorithm for the single and multiple terminal delivery problem, *Mgmt Sci.*, **18**, 664–82

Tomizawa, N. (1971), On some techniques useful for the solution of transportation network problems, *Networks*, **1**, 173–94

Traub, J.F. ed. (1976), *Algorithms and Complexity: New Directions and Recent Results* (Academic Press, New York)

Wagner, H.M. (1969), *Principles of Operations Research with Applications to Managerial Decisions* (Prentice-Hall, Englewood Cliffs, N.J.)

White, L.S. (1969), Shortest route models for allocation of inspection effort on a production line, *Mgmt Sci.*, **15**, 249–59

Wiest, J.D. (1967), A heuristic model for scheduling large projects with limited resources, *Mgmt Sci.*, **13**, B359–77

Wilson, R.J., and Beineke, L.W. (1979), *Applications of Graph Theory* (Academic Press, London)

Wren, A., and Holliday, A. (1972), Computer scheduling of vehicles from one or more depots to a number of delivery points, *Opl Res. Q.*, **23**, 333–44

Wright, J.W. (1975), Reallocation of housing by use of network analysis, *Opl. Res. Q.*, **26**, 253–8

Yao, A. Chi-Chih (1975), An $O(|E| \log \log |V|)$ algorithm for finding minimum spanning trees, *Inf. Process. Lett.*, **4**, 21–3

Yellow, P.C. (1970), A computational modification to the savings method of vehicle scheduling, *Opl Res. Q.*, **21**, 281–3

Yen, J.Y. (1971), Finding the k-shortest, loopless paths in a network, *Mgmt Sci.*, **17**, 712–15

Yen, J.Y. (1972), Finding the lengths of all shortest paths in N-node non-negative distance complete networks using $\frac{1}{2}N^3$ additions and N^3 comparisons, *J. Ass. comput. Mach.*, **19**, 423–4 [see also correction by Williams, T.A., and White, G.P. (1973), *J. Ass. comput. Mach.*, **20**, 389–90]

Zadeh, N. (1972), Theoretical efficiency of the Edmonds–Karp algorithm for computing maximal flows, *Communs Ass. comput. Mach.*, **19**, 184–92

Zangwill, W.I. (1969), *Nonlinear Programming: A Unified Approach* (Prentice-Hall, Englewood Cliffs, N.J.)

Zufryden, F.S. (1975), Media scheduling and solution approaches, *Opl Res. Q.*, **26**, 283–95

Index

activity   121
activity-on-vertex network   130
admissible   149
almost everywhere (ae)   270
alternating basis   250
AND–OR graph   273, 274
arc   1, 13
    backward   202
    forward   202
    reverse   218
    saturated   199
arc flow   196
arc-to-vertex dual   130
assignment problem (linear)   53, 154, 170, 250
assortment problem   41
augmented predecessor index method   244
augmented threaded list   246

B & B trees   50
    heuristic   265
backtracking   50
backward pass   126
behaviour of functions   282
    exponential   282
    factorial   282
    logarithmic   282
    polynomial   282
bi-directional search   74
bounding   53
bounding function   53
bounding rule   53
branching   49, 50
    exclusion   62
    immediate successor   51
    inclusion   62
    inclusion/exclusion   62
    rank   51
    successor   51

call set   175
capacity of arc   196
capacity of cut   201
catchment area   259
1-centdian   100
1-centdian problem (1-CDP)   100
1-centre problem (1-CP)   100
1-centre problem in the plane (1-CPP)   96
chain   17
    elementary   17
    Euler   19
    flow augmenting   202
    Hamilton   157
Chinese Postman Problem (CPP)   163
circuit   17
    Euler   19
    Hamilton   149
complexity of algorithms   282
components   18
consistent function   76
constraints, capacity   197
    conservation   197
    non-negativity   197
costs, direct and indirect   134
crash rate   138
crashing   136
critical path   126
Critical Path Method (CPM)   121
cut   137, 200
    associated   202
    capacity of   201
    crashable   138
    irreducible   204
    positive   138
    reduction of   204
cutoff   38
cycle   17
    absorbing   226
    active   226
    creating   226
    Euler   19
    Hamilton   149

Index

dead-end 209
degree 21
    in- 20
    out- 20
developing a node 50
development rule 54
    breadth-first (BF) 55
    change-when-necessary 55
    depth-first (DF) 54
    uniform-cost (UC) 55
deviation 87
dominance 56
dual, arc-to-vertex 130
duality gap 109, 114, 161
duration, crash 134
    estimated 123
    normal 134
dynamic programming 38

earliest finish time 127
earliest start time 127
edge 1, 14
edge bottleneck point 115
event 123

facilities location problem,
    emergency 94
    ordinary 94
    private sector 94
    public sector 94
facility 94
float, free 129
    independent 129
    safety 129
    total 127
flow, arc 196
    complete 200
    conserved 196, 197
    dynamic 222
    entering 224
    leaving 224
    minimal cost 228
    multi-commodity 227
    network 196, 197
    nonlinear 237
    with gains 224
flow shop sequencing problem 66, 171
forward pass 126
function, approximation 262
    bounding 53
    see also behaviour of functions

gain 224
globally optimal 257
graph 2
    AND–OR 274
    bipartite 18
    complete bipartite 18
    complete directed 16
    connected 18
    construction 262
    directed 13
    mixed 14
    multi- 43
    neighbourhood 256
    planar 15
    search 256
    sparse 28
    symmetric 13
    undirected 14

hill-climb 258

implicit enumeration 37
inactive node 54
incumbent solution 54
in-degree 20
intuitive method (for maximal flow) 199

job-splitting 140

$K_{m,n}$ 18
K_s, K_n 16
kilter, in 220
    out of 220
kilter number 220, 233
knapsack problem (KP) 39
Konigsberg Bridges Problem 43
Kuratowski's theorem 16

λ-opt tours 163
label 3
    permanent 72
    temporary 72
label correcting algorithm 83
label setting algorithm 83
latest finish time 127
latest start time 127
layer 208
length (of chain) 230
level 51
locally optimal 257
location set covering problem (LSCP) 111

loop-factor 248

media scheduling 45
makespan 66
matching 166, 167
matching set of chains 164
matrices associated with graphs 21
matrix, boolean adjacency 26
    reachability 27
maximal covering location problem (MCLP) 111, 112
1-median problem (1-MP) 96
1-median problem in the plane (1-MPP) 95
p-median problem (p-MP) 106
method, heuristic 255
    improvement 256
    nearest neighbour 266
    two-step (TS) 267
minimal spanning m-forest 35
minimal spanning tree (MST) 31

neighbour 256
    nearest (NN) 266
    3- 161
neighbourhood 256
    function 256
    graph 256
network 31
    activity-on-arc 130
    activity-on-vertex 130
    directed 31
    DK augmentation 208
    undirected 31
node 37
    active 54
    leaf 38
    root 37, 38
    terminal 38
NP 285
NP-complete 286
NP-hard 286

Open TSP 157
order 282
out-of-kilter algorithm 221, 222, 234

P 283
partition, proper 37
path 17
    elementary 17
    Euler 19

    Hamilton 27, 157
    superior 199
penalties 160
PERT 131, 132
pictures 14
planar 15
polynomially bounded 283
polynomially reduced 285
predecessor 13
primitive problem 272
Principle of Optimality 37
profile 140
project duration 126
project network 123
pseudo-cost 59
pseudo-depot 175

quasi-tree 247

random descent 258
reduction of matrix 149
resource constrained problem 141
resource levelling 141, 143
resource smoothing 141
reverse graph 13
root, of deviation 88
    of tree 37, 243

savings 181
SD-tree 72, 76
search, bi-directional 74
    uni-directional 74
seed 186
selection rule 51
set covering problem (SCP) 109, 110
simple location problem (SLP) 101
sink 71
source 71
spur 88
stage 38
start point 258
state 38
steepest descent 259
Steiner problem 30
strategy, bounding 53
    branching 50, 62
    climbing 258
    developing 54
    random descent 258
    steepest descent 259
subtour 155
successor 13

Index

super source (and sink)   214
SWEEP   186

terminal (node)   38
time–cost tradeoff   133
3-transformation   161
λ-transformation   163
transshipment problem (TRP)   240
travelling salesman problem (TSP)   148
tree   2, 29
   alternative definitions   30
   B & B (search)   50
   complete enumeration   38
   directed   72

M-alternating   167
minimal spanning (MST)   31
SD-   72, 76
search   37
spanning   31
Steiner   30

value (of flow)   197
vehicle routing problem (VRP)   174
vertex   1, 13
vertex number   233

weights (or arcs or edges)   31
worst case analysis   286